Holt Science and Technology
Grade 6

Tennessee Comprehensive Assessment Program Test Preparation Workbook

Printed in the United States of America

ISBN 13: 978-0-55-401784-6
ISBN 10: 0-55-401784-9

1 2 3 4 082 11 10 09 08

Contents

Introduction

This workbook consists of practice activities designed to prepare your students to take the *Tennessee Comprehensive Assessment Program Test*. The questions are correlated to the *Tennessee Science Standards* as well as the appropriate Skills and Processes Standards. This breadth of content coverage provides teachers with an opportunity to assess their students' understanding of the essential science knowledge and skills at the middle school level. These assessments can then help identify topics or concepts in need of re-teaching or additional practice and should be used to inform curricular decisions on the classroom or school levels.

Grade 6 – Inquiry

GLE 0607.Inq.1 Design and conduct open ended-scientific investigations.

STANDARD REVIEW

Designing a good experiment requires planning. Every factor should be considered. For example, a possible hypothesis about deformities in frogs is that they were exposed to an increased amount of ultraviolet (UV) light as eggs. If an increase in exposure to ultraviolet light is causing the deformities, then some frog eggs exposed to increasing amounts of UV light in a laboratory will develop into deformed frogs.

An experiment to test this hypothesis is summarized in Table 1. In this case, the variable is the length of time the eggs are exposed to UV light. All other factors, such as the temperature of the water, are the same in the control group and in the experimental groups.

In a well-designed experiment, the differences between control and experimental groups are caused by the variable and not by differences between individuals. The larger the groups are, the smaller the effect of a difference between individual frogs will be. The larger the groups are, the more likely it is that the variable is responsible for any changes and the more accurate the data collected are likely to be. Scientists test a result by repeating the experiment. If an experiment gives the same results each time, scientists are more certain about the variable's effect on the outcome.

Table 1 Experiment to Test Effect of UV Light on Frogs

Group	Control Factors			Variable
	Kind of frog	Number of eggs	Temperature	UV light exposure
1 (Control)	Leopard frog	100	25°C	0 days
2 (Experimental)	Leopard frog	100	25°C	15 days
3 (Experimental)	Leopard frog	100	25°C	24 days

GUIDED PRACTICE

Directions: Using the Standard Review and what you have studied, read each question and circle the letter of the best response.

Which term below best describes a hypothesis?

A the result of an experiment

B a well-tested explanation for why something happens

C a random guess about what causes something

D a possible explanation of an observation or result

The correct answer is D. A hypothesis is a possible explanation based on observations or reason, so it is not a random guess (Answer C). It is the basis for an experiment, not a result (Answer A). A well-tested explanation (Answer B) is a theory, not a hypothesis.

STANDARD PRACTICE

1. What can you conclude if you perform one test and the results support the hypothesis with which you started?

- **A** The hypothesis is correct.
- **B** The data were recorded incorrectly.
- **C** More experiments are needed to confirm the result.
- **D** The hypothesis has become a theory.

2. Lisa made a hypothesis that gravity causes all things to fall to the ground. Her sister pointed out that helium balloons move away from the ground. Lisa should

- **F** throw out her hypothesis and do a different research project.
- **G** try to find out why helium balloons are different from most objects.
- **H** ignore her sister and keep testing her own hypothesis.
- **J** assume that gravity does not affect helium balloons.

3. How could you test the hypothesis that marigolds need more water than cacti?

- **A** Plant marigolds in the desert and observe whether the seeds form plants.
- **B** Give the same amount of water to each type of plant and observe their growth.
- **C** Stop watering a cactus plant and observe whether it dies.
- **D** Observe the differences between plants grown in a warm place and in a cool place.

4. a. Why is a control important in an experiment to test a hypothesis?

b. How can a hypothesis be useful even if it is proved incorrect by experiments?

Grade 6 – Inquiry

GLE 0607.Inq.2 Use appropriate tools and techniques to gather, organize, analyze, and interpret data.

STANDARD REVIEW

In most experiments, you will measure some quantity, such as distance, mass, temperature, or time. When you work in the lab, you will use scientific apparatus to make these measurements and control experiments. It is important that you follow the correct procedures for using the apparatus. Techniques for using equipment will be provided by your teacher or in instructions in your lab book. It is important to read and follow all procedures exactly. If you use equipment incorrectly, you may create a safety hazard or get incorrect results. It is important to follow instruction in the correct sequence and observe all safety guidelines.

After you finish collecting data, you will use tables, graphs, diagrams, maps, or other visual displays, to analyze and interpret it. These tools point out the relationships in the numbers that record your measurements. Part of conducting a successful experiment is analyzing your data to find any hidden patterns. Two common patterns that you might see on a graph of experimental data are *linear relationships*, in which data tends to form a straight line, and *repeating relationships*, in which a cycle of changing values shows up in the analysis. When you analyze data you also need to look at methods of measurement and calculation to find the difference between a predicted amount and a calculated amount.

GUIDED PRACTICE

Directions: Using the Standard Review and what you have studied, read each question and circle the letter of the best response.

What should you do if you are not sure how to use a piece of laboratory equipment?

A Stop working on the experiment.

B Ask another student to show you how to use the equipment.

C Check the lab instructions for the correct techniques.

D Make your best guess about how it works.

The correct answer is C. When you run an experiment that uses equipment, you can get information on using it from the lab instructions or from your teacher. You should find out how to use the equipment and then continue the experiment (Answer A). You should never guess if you don't know how to use the equipment (Answer D) or ask someone other than your teacher (Answer B) because you might introduce an error that could cause a safety hazard or change the results of the experiment.

Grade 6 – Inquiry

STANDARD PRACTICE

1. Which of the following instructions would not be useful in an instruction sheet for a lab experiment?

A Turn the first dial slowly to the right until the meter reads 3.0.

B Do not begin until your teacher checks the apparatus.

C Add water until the liquid reaches the 50 mL mark.

D Put some oil in the beaker but be careful not to add too much.

2. As part of a field investigation on the thermal pollution of waterways, a scientist needs to measure the temperature of a nearby river. Which of the following pieces of equipment would be the best choice for making the measurement?

F barometer

G thermometer

H manometer

J speedometer

3. Which of these statements is the best description of when you should wear eye protection in the laboratory?

A when your teacher says protection is necessary

B when there is even a very small chance of harm to your eyes

C when you are using an instrument that has moving parts

D when your lab partner reminds you to put on your goggles

4. a. A procedure for using a balance to measure the mass of a liquid says to measure the mass of an empty beaker, and then add the liquid to the beaker. Why do you need to follow the procedure in that order?

b. What type of equipment would you use to measure the mass of the beaker?

GLE 0607.Inq.3 Synthesize information to determine cause and effect relationships between evidence and explanations.

STANDARD REVIEW

Once scientists finish their tests, they must analyze the results. Scientists often make graphs and tables to organize and summarize their data. Analysis also includes comparing new data with known information to discover what new information the data provides. After carefully analyzing the results of their tests, scientists must decide whether their results supported the hypothesis. The conclusion is an interpretation of the results and how they compare to the original ideas of the hypothesis. If a scientist concludes that the results support the original hypothesis, the conclusion may suggest new questions for further study.

One way to test a hypothesis is to do a *controlled experiment.* A controlled experiment compares the results from a control group with the results from one or more experimental groups. The control group and the experimental groups are the same except for one factor. This factor is called a *variable.* The experiment will then show the effect of the variable. If your experiment has more than one variable, determining which variable is responsible for the experiment's results will be difficult or impossible.

GUIDED PRACTICE

Directions: Using the Standard Review and what you have studied, read each question and circle the letter of the best response.

In an experiment, a closed container is heated and the pressure inside the container is measured. Which of these conclusions can you make based on the data below?

Pressure measured in a closed container			
Volume	Temperature	Gas in Container	Pressure
100 mL	0°C	nitrogen	100 kPa
100 mL	25°C	nitrogen	109 kPa
100 mL	50°C	nitrogen	118 kPa

A Increasing the volume causes an increase in pressure.

B Increasing the temperature of nitrogen in a closed container causes the pressure to increase.

C Increasing the temperature of any material causes its pressure to increase.

D Increasing the pressure in a closed container causes the temperature to increase.

The correct answer is B. Answer D is incorrect because the experiment changed the temperature and observed the pressure. Answer A is incorrect because the volume was constant. Because only nitrogen was studied, you cannot make conclusions about other materials (Answer C).

Grade 6 – Inquiry

STANDARD PRACTICE

1. Based on the information in the table below, which of the following is <u>not</u> a valid conclusion about carbon dioxide emissions?

U.S. Carbon Dioxide Emissions* from Fossil Fuel Energy Consumption			
Energy Sector	1997	1998	1999
Residential	289	289	290
Commercial	241	244	244
Industrial	490	480	481
Transportation	474	482	496
Total	1493	1495	1511

*Emissions are given in millions of metric tons of carbon.

A Carbon dioxide emissions increased each year from 1997 to 1999.

B The commercial energy sector is the smallest source of emissions.

C Carbon dioxide emissions have always exceeded 1400 million metric tons.

D The transportation and industrial energy sectors are the largest sources of carbon dioxide emissions.

2. How might a disease that killed the grasses in an area affect animals such as coyotes and mountain lions, which do not eat grass?

F decrease the number of animals on which they feed

G increase the number of other predators that compete with them

H no effect because coyotes and mountain lions don't eat grass

J increase the food supply because small animals can no longer hide

3. This graph shows the levels of carbon dioxide (CO_2) in the atmosphere from 1860 through 1980.

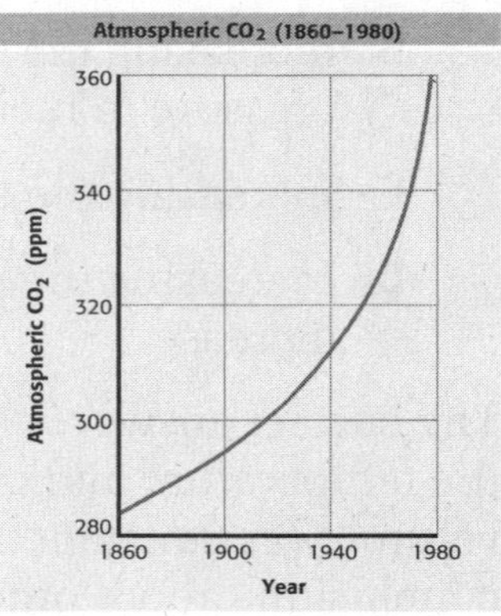

a. Was the rate of change in the level of CO_2 between 1940 and 1960 higher or lower than it was between 1880 and 1900? How can you tell?

b. What conclusions can you draw from reading this graph?

GLE 0607.Inq.4 Recognize possible sources of bias and error, alternative explanations, and questions for further exploration.

STANDARD REVIEW

Established theories are usually built on many experiments and observations. When new experimental results do not agree with the existing theory, scientists generally do not revise the theory immediately. Sometimes the results of the experiment contains an error in the data or the analysis. Even after an experiment has been reproduced independently, more information may be needed. Different scientists may interpret the results in different ways. In that case, further experiments are needed. The original experiment might suggest new ways to test the hypothesis.

Scientific investigations are a continual process. Even after results are reviewed and accepted by the scientific community for publication, the investigation of the topic may not be finished. New evidence may become available. The scientist may change the hypothesis based on the new evidence. In other cases, the scientist may have more questions that arise from the original evidence. When a number of different experiments all provide data that are not consistent with the theory, the theory is usually revised.

An example of a hypothesis that was not immediately accepted was Alfred Wegener's hypothesis of continental drift. Wegener analyzed fossils on several continents, and he found that some continents had fossils similar to fossils on other continents. After making many observations, Wegener proposed that the continents were once a supercontinent, but later drifted apart. Although his results were confirmed by other scientists, Wegener's hypothesis was not generally accepted right away. Many scientists did not accept his hypothesis because it did not seem possible for the crust to move in the way he proposed. Until further investigations showed how the continents could move, continental drift was not generally accepted. Wegener's observations are now part of the support for the theory of *plate tectonics*, which describes the movement of the continents.

GUIDED PRACTICE

Directions: Using the Standard Review and what you have studied, read each question and circle the letter of the best response.

A report of which of the following observations would be inconsistent with current theories about Earth?

- **A** Satellite data that shows continents moving.
- **B** Measurements show Earth's core is made of molten rock.
- **C** Earth's revolution around the Sun takes slightly more than 365 days.
- **D** Observations show air pressure is lowest at sea level.

The correct answer is D. There is no current theory that would explain air pressure being lowest at sea level. The other observations are all consistent with our current understanding of how the planet functions.

STANDARD PRACTICE

1. Which of the following is not a reason for scientists to wait to revise a theory based on inconsistent results? Scientists must

A meet to discuss the change.

B carefully consider the new data to be certain they are not flawed.

C see a significant difference in order to change theories based on many experiments.

D confirm and support the new data by other evidence.

2. During a study at the South Pole, researchers record a sudden increase in the intensity of solar radiation reaching the Pole. No increase was observed at the equator. Which of the following is a reasonable reaction to the results?

F Conclude that something has changed within the sun.

G Conclude that the measurements at the equator are in error.

H Collect additional data with a different instrument for comparison.

J Revise theories of how sunlight reaches Earth.

3. Which of these observations should be considered to be an experimental error?

A Observation of a lunar eclipse during the moon's first quarter.

B Data indicating that a major ocean current has changed direction.

C A satellite measurement showing a small increase in the height of a mountain.

D An observation that rain is falling while the sun is shining.

4. a. How do scientists determine whether new results are significant?

b. Why must researchers consider the precision of their instruments when they are determining the significance of data?

Grade 6 – Inquiry

GLE 0607.Inq.5 Communicate scientific understanding using descriptions, explanations, and models.

STANDARD REVIEW

A key part of scientific research is communicating the results of an experiment. Scientific writing must present detailed information in a way that is very specific. That means it is very different from the literary writing that is used to tell a story or support a political point of view. Whether you are writing a laboratory report for your teacher or submitting a paper to a scientific journal, your lab report should contain enough information so that others can use it to reproduce your experiment and compare their results to yours.

A pattern, plan, representation, or description designed to show the structure or workings of an object, system, or concept is a *model.* With a model, a scientist can explain or analyze an object, system, or concept in more detail. Models are used in science to help explain how something works or to describe how something is structured.

After scientists have collected their data, they organize it into tables, charts, or graphs in order to look for trends and identify relationships among the variables. These visuals can show new information. For example, all of the data in a single row or column in a table have at least one characteristic in common. That characteristic is labeled in the row or column's label. In the table below, all of the values in the first column are temperatures given in degrees Celsius.

Temperature (°C)	Time to double bacteria population (min)
10	130
20	60
30	29
40	19
50	no growth

GUIDED PRACTICE

Directions: Using the Standard Review and what you have studied, read each question and circle the letter of the best response.

How does scientific writing differ from literary writing?

A The two types of writing have different purposes.

B Scientific reports are always written in English.

C Science writing is shorter than literary writing.

D Science writing is about facts, and literary writing is fiction.

The correct answer is A. Science writing must present details in a clear and detailed way, unlike literary writing. Scientific reports can be written in any language (Answer B), and they can be long or short, depending on the amount of information to be presented (Answer C). Nonfiction literary writing is also about facts (Answer D).

STANDARD PRACTICE

1. The main purpose of scientific writing is to

A entertain the reader.

B convince other people that a hypothesis is true.

C report information or data.

D get credit for a new discovery.

2. According to the graph of population growth below, the world population in the year 2050 will be

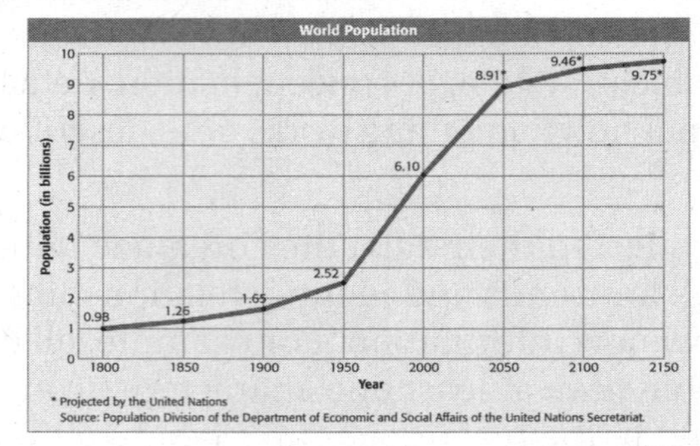

F 891.

G 891,000,000.

H 8,910,000,000.

J 89,100,000,000.

3. Rhonda observed that plants growing on the windowsill grew faster than similar plants growing in a cool, shady corner of the room. How should she communicate her observation in her lab report?

A. Plants grew faster when they were on a windowsill

B. Some plants grew faster than other plants

C. The plants in a warm, sunny location grew faster than plants in a cool, shady location

D. The growth rate of plants seems to depend on light and temperature

4. a. Why do scientific reports include a detailed description of how the experiment was performed?

b. For what types of scientific investigations should a report be written?

Grade 6 – Technology and Engineering

GLE 0607.T/E.1 Explore how technology responds to social, political, and economic needs.

STANDARD REVIEW

Scientists study the natural world. Engineers work to put scientific knowledge to practical use and to build the tools to use scientific knowledge. Some engineers design and build the buildings, roads, and bridges that make up cities. Others design and build electronic things, such as computers and televisions. Some even design processes and equipment to make chemicals and medicines. Engineers may work for universities, governments, and private companies.

While the driving force behind science is curiosity, the driving force behind technology is finding a solution to a social, political, or economic need. Technologies that societies need include power, communication, medical care, and transportation. Science and engineering together address these needs as they continually change. The internet, for example, combines many technologies to link distant parts of the world together in ways that were impossible even a few decades ago. Each communication advance leading to the internet combined scientific knowledge and engineering application. Political needs in a society include infrastructure used by everyone, such as roads and bridges. Another political need is defense, leading to the development of military technologies. Economic needs drive many technologies, including the development of better systems of manufacturing and distribution of materials.

Technology is a very broad term that includes many different types of application of science. It can be an object or device, such as a medical imaging machine. It can be a technique or method, such as a new way to make a fuel from grain crops. Technology can also be a system of production, for example, the assembly line, which changed how products were made throughout the world.

GUIDED PRACTICE

Directions: Using the Standard Review and what you have studied, read each question and circle the letter of the best response.

Which of the following is an example of technology?

The correct answer is D. All of these are examples of different types of technology. A pair of pliers (Answer A) is not a newly discovered technology, but it is the application of scientific principles to make a tool. The technique describe in Answer B is an example of a new method of doing something based on science. Robotic manufacturing (Answer C) is a technological system that has changed many manufacturing industries.

Grade 6 – Technology and Engineering

STANDARD PRACTICE

1. Which of these scientific principles had to be studied before weather forecasting technologies could be developed?

A how changes in the atmosphere affect precipitation
B the distance between Earth and the sun
C how raindrops create rainbows
D pollution of the atmosphere by burning fossil fuels

2. Which technology was developed by scientists and engineers who studied the properties of charged particles?

F new types of ceramic glazes
G internal combustion engine
H transmission of electrical energy
J better alloys for airplane engines

3. In what way could a pesticide that kills an invasive species, but not a natural species in an environment be considered a new technology?

A It is more effective than other pesticides.
B It is a practical application of science.
C It is done first in a laboratory.
D It is profitable for the pesticide company.

4. a. In what way does the internet represent a technology that meets an economic need of society?

b. What are some of the scientific principles that led to the development of the internet?

Grade 6 – Technology and Engineering

GLE 0607.T/E.2 Know that the engineering design process involves an ongoing series of events that incorporate design constraints, model building, testing, evaluating, modifying, and retesting.

STANDARD REVIEW

Technology is the process by which humans modify nature to meet their needs. Technology includes the products, equipment, and systems that you use every day, but it is more than that. Technology also includes the processes used to develop and build those products.

A major part of developing new technologies is engineering design. *Engineering* consists of the knowledge of the design to make products and develop processes to solve problems. As with scientists, the work of engineers must follow the laws of nature. Engineers also have to consider what materials are available, what safety problems may exist, and what the effects are on the environment.

In order to achieve their goals and develop new technologies, engineers use a design process. This process has many steps but it can be summarized as follows:

- identify the problem to be solved
- conduct research
- make decisions about materials and processes
- design and build models
- perform tests and evaluate the results
- modify the product or system
- test again and repeat the design process as many times as necessary
- build the technology

During this process, engineers may try many different materials and arrangements. When the change makes the product better, less expensive, more reliable or improved in any other way, it may be included in the next test. New scientific discoveries can become part of new design, making designs work even better.

GUIDED PRACTICE

Directions: Using the Standard Review and what you have studied, read each question and circle the letter of the best response.

As part of an engineering design process, engineers want to make a less expensive disposable towel. Which of these materials would they most likely want to test?

A silk **B** iron

C paper **D** fiberglass

The correct answer is C. Of the choices, paper is both the least expensive and most likely to work as a towel. Silk (Answer A) is a fabric that could be used, but it is much too expensive for a disposable product. Iron and fiberglass (Answers B and D) can both be woven into a flexible, cloth-like material but it would not be likely to work as a towel. Both materials would also be too expensive for a disposable towel even if they could work for the purpose.

Grade 6 – Technology and Engineering

STANDARD PRACTICE

1. Which of the following would be most important in designing a new drug to treat headaches?

A low cost of each pill

B making sure the drug does not have harmful side effects

C size of the pill

D making sure the drug does not taste bad

2. An engineer wants to design a sensor that can be released in the ocean to monitor ocean currents and send signals to satellites. Which of these would not be an important design problem?

F The sensor corrodes when it is exposed to salt water.

G The plastic housing for the sensor is only available in a few colors.

H The battery runs out of charge after two days.

J The unit has small leaks that cause it to fill with water and sink.

3. a. Why would an engineer designing a new type of boat start the engineering design process by using a small model instead of a full-scale model?

b. Many modern boats are made of fiberglass. Why might engineers try designs using fiberglass instead of metal or wood?

Grade 6 – Technology and Engineering

GLE 0607.T/E.3 Compare the intended benefits with the unintended consequences of a new technology.

STANDARD REVIEW

The purpose of technology is to use science and engineering to solve problems and provide products and processes that people need. Technology has provided all the systems that hold our society together, including electric power, transportation networks, medical care, food supplies, and many others. Unfortunately, technology systems do not always work exactly as they are planned. Sometimes systems break down and stop working. Often, in addition to the planned benefits of technology, there are undesirable effects. These *unintended consequences* are problems caused by technology, which were not expected.

An example of an unintended consequence of technology is air pollution caused by burning fossil fuels. Transportation technology has certainly served its purpose. Automobiles, trucks, and airplanes provide the intended benefit, allowing people and goods to move easily from one place to another. At the same time, there have been unintended consequences of this technology. Acid rain caused by impurities in gasoline has damaged structures and destroyed stream ecosystems. Carbon dioxide released into the atmosphere when fuels burn is expected to cause dramatic changes in the world's climates. Another unintended consequence of transportation technology based on oil is political tension, which is the result of the uneven distribution of oil resources among different countries.

GUIDED PRACTICE

Directions: Using the Standard Review and what you have studied, read each question and circle the letter of the best response.

Which of these is an unintended consequence of technology?

- **A** communicating with people in other countries by instant messages
- **B** using your computer to find information about an illness from a library far from you home
- **C** people finding your personal information by using their computers
- **D** being able to purchase a CD online that is not available in a local store

The correct answer is C. One of the unintended consequences of the internet is that people can sometimes find information online that should not be public. The other choices are all intended benefits of internet technology: the ability to communicate easily (Answer A), find information quickly (Answer B) and shop at stores in other cities (Answer D).

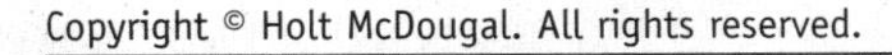

STANDARD PRACTICE

1. What could be an unintended consequence of building wind turbines on top of a mountain?

- **A** increase in the amount of electricity available in a nearby town
- **B** reduction in the cost of producing power
- **C** soil erosion when a road is built to get the turbines up the mountain
- **D** decrease in emission of greenhouse gases

2. Which of the following is an intended benefit of a new technology?

- **F** A new type of weather radar improves the ability to predict tornadoes.
- **G** A pesticide kills all of the honeybees in an orchard.
- **H** Traffic backs up on all the roads near a newly built bridge.
- **J** People using cell phones while driving cause an increase in highway fatalities.

3. Which of these effects is an unintended consequence of technology?

- **A** A new locomotive design reduces travel time by one hour.
- **B** Changing some operations increases production in a manufacturing plant.
- **C** Replacing the materials in a truck body reduces rust damage.
- **D** A software upgrade causes a computer to stop working.

4. a. What are some of the intended benefits of a dam built to produce hydroelectric power?

b. What are some possible unintended consequences of a dam built to produce hydroelectric power?

Grade 6 – Technology and Engineering

GLE 0607.T/E.4 Describe and explain adaptive and assistive bioengineered products.

STANDARD REVIEW

Many technologies are based on new discoveries about life and how organisms change and adapt to their environments. *Bioengineering* applies the ideas of engineering to the science of living things. Chemical processes inside living cells control life and death, growth, reproduction, and how the cell interacts with its environment.

Scientists and engineers study the chemical reactions inside cells and learn how cells adapt to their environment. They also look at the structure of tissues and organs in plants and animals.

By using what they learn about living things, bioengineers develop new, useful products. Many of these products are used for health care. They use the knowledge of how cells operate to design new drugs that work better. Materials that can be implanted inside the human body are used to make artificial joints. Other new materials can replace body tissues, such as heart valves that are damaged by a viral infection. New bioengineered systems have allowed blind people to see and deaf people to hear by connecting devices directly to the body's nervous system. Bioengineers have also developed artificial arms and legs that can be controlled, through the nervous system, by a person's brain.

Bioengineers also work in other fields. In agriculture, for example, they modify the genes inside the cells of plants so that the plants kill harmful insects that eat their leaves. Other plants have been modified to produce bigger crop yields or to increase the amount of oil in their tissues, which can be extracted for use in fuel. These fuel oils from bioengineered plants are called *biofuels*.

In some cases, bioengineers can actually develop organisms to produce a material that is useful or performs a desirable function. Insulin, a medicine for treating diabetes, can be manufactured by bioengineered bacteria and then purified for medicinal use. Some bioengineered bacteria eat chemicals that are spilled in the environment. These bacteria have been modified so that they survive by eating oil or other chemicals absorbed by the ground or floating in water.

GUIDED PRACTICE

Directions: Using the Standard Review and what you have studied, read each question and circle the letter of the best response.

Which of these is the best example of a product developed by bioengineers?

A the internet

B cars that use less fuel

C bacteria that produce silk

D a new metal alloy

The correct answer is C. Although it is possible that bioengineering could have been used in the other products, the design of a new living cell is the best example of bioengineering because it modifies a living cell. The internet (Answer A) is based on electronics and system engineering, fuel-efficient cars (Answer B) are generally developed by mechanical engineering, and metal alloys are developed by chemical engineering.

Grade 6 – Technology and Engineering

STANDARD PRACTICE

1. **How could the designer of a communication system use bioengineering to improve reliability?**

A She could apply the design of the network of nerves in the body to their system connections.

B She could study the materials of bones to find a better electrical conductor.

C She could develop new bacteria to carry messages on the system.

D Bioengineering cannot be applied to communication systems because they are not living things.

2. **How is bioengineering applied to the development of new drugs?**

F making advertisements that tell doctors about the drugs

G finding plants that contain natural drugs

H reducing costs by building a larger manufacturing plant

J making drugs based on knowledge about the chemical reactions in a cell

3. **a. Why are biological systems useful for designing other kinds of systems?**

b. In what way is bioengineering a form of technology?

Grade 6 – Life Science: Standard 2 – Interdependence

GLE 0607.2.1 Examine the roles of consumers, producers, and decomposers in a biological community.

STANDARD REVIEW

Organisms that change the energy in sunlight into chemical energy, or food, are called *producers.* They do this by using a process called *photosynthesis.* Most producers are plants, but algae and some bacteria are also producers. Not all organisms can make their own food in the way that producers can. Organisms that eat other organisms are called *consumers.* Consumers eat producers or other animals to obtain energy.

Organisms that get energy by breaking down dead organisms are called *decomposers.* Bacteria and fungi are decomposers. These organisms remove stored energy from dead organisms. They produce simple materials, such as water and carbon dioxide, which can be used by other living things. Decomposers are nature's recyclers. All living things are connected in the web of life. In the web of life, energy and other resources flow between organisms and their environment.

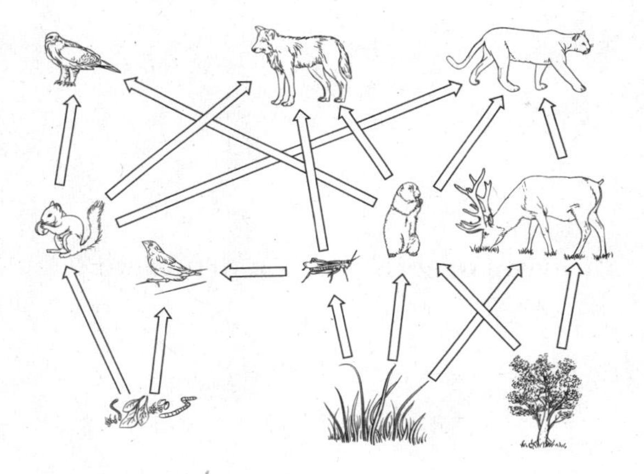

GUIDED PRACTICE

Directions: Using the Standard Review and what you have studied, read each question and circle the letter of the best response.

Nature's recyclers are

A predators.

B decomposers.

C producers.

D omnivores.

The correct answer is B. A decomposer is an organism that obtains energy by breaking down other organisms after they die, making their resources available to other organisms. Producers (Answer C) obtain energy from sunlight. Predators (Answer A) obtain energy by eating living organisms, and omnivores (Answer D) obtain energy by eating plants and animals. Predators and omnivores are both types of consumers.

Grade 6 – Life Science: Standard 2 – Interdependence

STANDARD PRACTICE

1. The arrows on the food web on the previous page show that

A prairie dogs eat grass.

B deer eat prairie dogs.

C squirrels eat grass.

D squirrels eat coyotes.

2. Which of the following organisms is a consumer?

F hawk

G grasses

H algae

J bacteria

3. Which of the following organisms is a producer?

A mouse

B owl

C mushroom

D grass

4. Katrina is eating lunch. She has a turkey sandwich with lettuce and tomatoes on wheat bread.

a. List the consumers and producers in the scenario described above.

b. Katrina gained energy to live by eating her lunch. Where did this energy come from originally? Explain your answer.

Grade 6 – Life Science:
Standard 2 – Interdependence

GLE 0607.2.2 Describe how matter and energy are transferred through an ecosystem.

STANDARD REVIEW

When animals eat plants and other animals, matter and energy are transferred from one organism to another. A *food chain* is a diagram that shows how energy in food flows from one organism to another. Because few organisms eat just one kind of food, simple food chains are rare. The energy connections in nature are more accurately shown by a food web. A *food web* is a diagram that shows the feeding relationships between organisms in an ecosystem. Energy moves from one organism to the next in a one-way direction.

The energy at each level of the food chain can be seen in an energy pyramid. An energy pyramid is a diagram that shows an ecosystem's loss of energy. The energy pyramid has a large base and a small top. Less energy is available at higher levels because only energy stored in the tissues of an organism can be transferred to the next level.

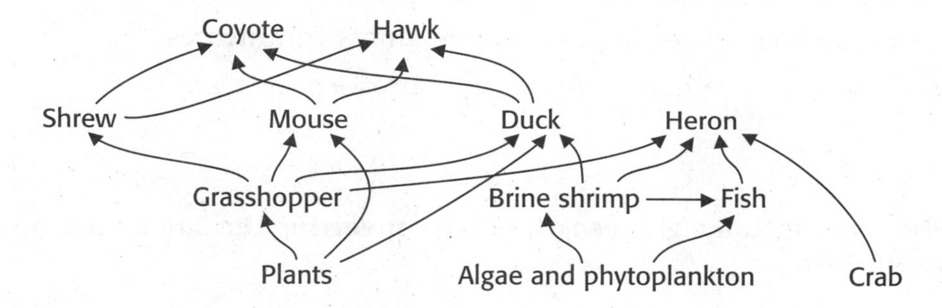

GUIDED PRACTICE

Directions: Using the Standard Review and what you have studied, read each question and circle the letter of the best response.

Which of the following statements best describes a food web?

A Many individual organisms of the same species live in the same space and share resources.

B An ecosystem is made up of a community of organisms and their environment.

C A black bear eats fruit and then spreads the fruit's seeds through its excretions.

D All life is connected by the transfer of energy and nutrients among organisms and their environments.

The correct answer is D. The food web shows the flow of energy within a system. The web contains more information than just the organisms in the system (Answer A) or one particular interaction (Answer C). It does not include information about the environment in which the members of the web live (Answer B).

STANDARD PRACTICE

1. The process by which energy moves through an ecosystem can be represented by

A food pyramids.
B energy cycles.
C food webs.
D photosynthesis reactions.

2. Which of the following is the correct order in a food chain?

F sun→producers→herbivores→scavengers→carnivores
G sun→consumers→predators→parasites→hosts
H sun→producers→decomposers→consumers→omnivores
J sun→producers→ herbivores → carnivores → scavengers

3. When plants make glucose, they are converting the sun's radiant energy into

A carbon dioxide and water.
B chemical energy that can be stored.
C producers and consumers.
D chlorophyll.

4. An energy pyramid is wide at the bottom and narrow at the top. Producers such as grasses and other plants are found at the bottom of the pyramid. Carnivores or scavengers are often found at the top of the pyramid.

a. What kind of organisms would you expect to find in the middle of the pyramid?

b. What two things do the widths of the different levels of the pyramid represent?

Grade 6 – Life Science: Standard 2 – Interdependence

GLE 0607.2.3 Draw conclusions from data about interactions between the biotic and abiotic elements of a particular environment.

STANDARD REVIEW

An organism's *environment* consists of all the things that affect it. These things can be divided into two groups, biotic and abiotic. All the organisms that live together and interact with one another make up the *biotic* part of the environment. The *abiotic* part of the environment consists of the nonliving factors such as water, soil, light, and temperature. In any environment, each organism is part of a *population,* or a group of individuals of the same species that live together. Organisms often compete for food, living space, and mates. A resource that is so scarce that it limits the size of a population is called a *limiting factor.* Any single biotic or abiotic resource can be a limiting factor to a population's size.

Sunlight and water are two abiotic factors that affect organisms in any environment. For example, most plants need sunlight for photosynthesis and water to grow. However, plants in different environments need different amounts of sunlight and water. For example, a cactus can survive in the hot sun with little water. A sunflower cannot survive in the same environment as the cactus.

GUIDED PRACTICE

Directions: Using the Standard Review and what you have studied, read each question and circle the letter of the best response.

Because resources are in limited supply in the environment, their use by one individual or population

A can be easily replaced by photosynthesis.

B has no effect on the populations.

C increases the number of members in each population.

D decreases the amount available to other organisms.

The correct answer is D. No environment has unlimited resources, so if one organism or group of organisms uses a particular resource, it is not available to other organisms. This decrease is likely to cause the number of organisms in other populations to decrease, so Answers B and C are incorrect. Resources cannot be easily replaced by photosynthesis (Answer A) because photosynthesis is a process that depends on the availability of resources.

Grade 6 – Life Science: Standard 2 – Interdependence

STANDARD PRACTICE

1. An abiotic factor in marine ecosystems is

A the temperature of the water.
B the depth of the water.
C the amount of sunlight that passes through the water.
D all of the above.

2. What term describes all nonliving factors in an environment?

F abiotic
G biotic
H zoological
J biological

3. Drought conditions cause a limited amount of food for a deer population. What will likely happen to the deer?

A The deer will continue to reproduce successfully.
B The deer will not find enough food to survive in their current population.
C The deer will continue to find enough food to survive in spite of the drought.
D The deer will compete for the limited food and survive.

4. The correct term for two or more individuals or populations trying to use the same resource, such as food, water, shelter, space, or sunlight, is

F overpopulation.
G carrying capacity.
H competition.
J limiting factor

5. In rural Tennessee, the white-tailed deer population has increased due to high rainfall and the resulting growth and availability of plant food resources. In your own words,

a. Predict what will happen to deer and plant populations if drought conditions occur due to the lack of rain.

b. Predict what will happen to deer and plant populations if rainfall continues to be high.

Grade 6 – Life Science: Standard 2 – Interdependence

GLE 0607.2.4 Analyze the environments and the interdependence among organisms found in the world's major biomes.

STANDARD REVIEW

A biome is made up of many ecosystems. A *biome* is an area where the climate typically determines the plant community, which supports an animal community. Similar biomes are found in different parts of the world where the climate is similar. For example, the chaparral biome in California is similar to the Mediterranean scrub biome around the Mediterranean Sea.

All biomes and ecosystems have producers, consumers, and decomposers. For example, deer in South America and kangaroos in Australia are herbivores. Deer and kangaroos feed on the plants in their environment. When these organisms die, they will be decomposed by the bacteria and fungi in the environment.

BIOMES

Biome	Primary Vegetation	Annual Precipitation	Animal Life
Desert	Plants	<25 cm	Jackrabbits
Savanna	Grasses	90–150 cm	Herds of Grazing Animals
Taiga	Trees	35–75 cm	Moose
Temperate Deciduous Forest	Trees	75–250 cm	Deer, Bears
Temperate Grassland	Grasses	25–75 cm	Antelope
Tropical Rain Forest	Plants, Trees	200–450 cm	Diverse
Tundra	Mosses	<25 cm	Reindeer, Caribou

GUIDED PRACTICE

Directions: Using the Standard Review and what you have studied, read each question and circle the letter of the best response.

Which organism has the same role as lions in the savanna?

A algae in the ocean
B coyote in the chaparral
C grasshopper in the prairie
D tree frog in the tropical rain forest

The correct answer is B. In a savannah biome, the lion is a consumer that is the top predator in the ecosystem. The coyote has the same role in the chaparral. The algae in the ocean (Answer A) is a producer, while the grasshopper in the prairie and the tree frog in the rain forest (Answers C and D) are consumers that are also prey for larger predators.

STANDARD PRACTICE

1. In which biome would you find most nocturnal animals?

A desert

B taiga

C savanna

D tundra

2. Look at the map of Earth's climate zones below. Where are biomes with the largest number of plant and animal species located?

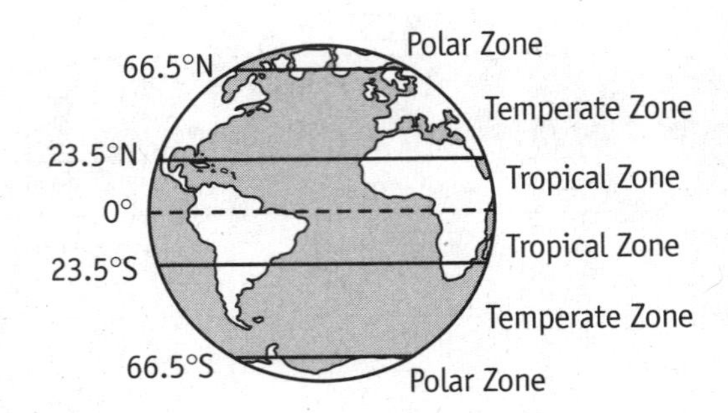

F in the polar zones

G in the tropical zones

H in the temperate zones

J only in the northern portion of the tropical zones

3. While excavating an area in the desert, a scientist discovers the fossils of very large trees and ferns.

a. What might the scientist conclude about biomes in this area?

b. What change in resource would most likely account for the difference in types of life in this area in the past compared to the current ecosystem?

Grade 6 – Earth and Space Science: Standard 6 – The Universe

GLE 0607.6.1 Analyze information about the major components of the universe.

STANDARD REVIEW

It is difficult to tell what type of galaxy we live in because the gas, dust, and stars keep astronomers from having a good view of our galaxy. Observing other galaxies and making measurements inside our galaxy, the Milky Way, has led astronomers to think that our solar system is in a spiral galaxy. The same elements, forces, and energy relationships that occur in the Milky Way appear to exist in other types of galaxies.

From our home on Earth, the universe stretches out farther than astronomers can see with their most advanced instruments. The universe contains a variety of objects. However, these objects in the universe are not simply scattered through the universe in a random pattern. The universe has a structure that is loosely repeated over and over again.

Measurements have shown that most galaxies are moving away from one another. Cosmologists think that the entire universe began with a single tremendous explosion. According to this theory—called the *Big Bang*—all of the universe was compressed under extreme pressure, temperature, and density into a very tiny spot. Then, billions of years ago, the universe rapidly expanded ("exploded"), and matter began to come together and form galaxies.

GUIDED PRACTICE

Directions: Using the Standard Review and what you have studied, read each question and circle the letter of the best response.

In the 1920s, Edwin Hubble discovered that

A the planets stay in orbit because of gravity.

B Earth orbits the sun.

C a telescope can be used to view objects in space.

D there are galaxies beyond the Milky Way.

The correct answer is D. Hubble discovered that indistinct patches that appeared to be clouds of gas were actually very distant clusters of stars. The other discoveries were made hundreds of years ago by Newton (Answer A), Copernicus (Answer B), and Galileo (Answer C).

Grade 6 – Earth and Space Science: Standard 6 – The Universe

STANDARD PRACTICE

1. Which term is used to describe the shape of the Milky Way galaxy?

A spiral
B elliptical
C globular
D irregular

2. In what part of the Milky Way galaxy is our solar system located?

F near the center
G above the central axis
H in a spiral arm between the center and edge of the galaxy
J at the outside edge of the spiral arms

3. Which of the following numbers is the best estimate of the number of stars in the Milky Way galaxy?

A 20,000,000
B 200,000,000
C 2,000,000,000
D 200,000,000,000

4. This graph shows Hubble's law, which relates how far galaxies are from Earth and how fast they are moving away from Earth.

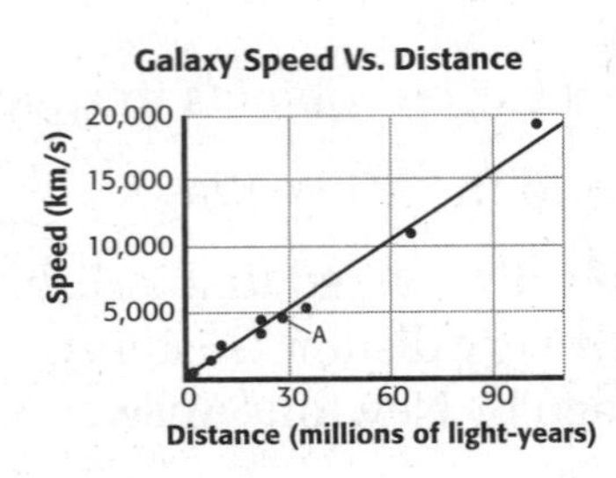

a. How far is galaxy A from Earth, and how fast is it moving away from Earth?

b. What does Hubble's Law state about the universe?

Grade 6 – Earth and Space Science: Standard 6 – The Universe

GLE 0607.6.2 Describe the relative distance of objects in the solar system from earth.

STANDARD REVIEW

The *inner solar system* contains the four planets that are closest to the sun. Mercury is the smallest of the planets and closest to the sun. Venus, like Mercury, orbits closer to the sun than does Earth. Earth and the moon orbit the sun together. The moon is much closer to Earth than any of the other major bodies in the solar system. Mars is a fourth planet of the inner solar system.

The *outer solar system* contains the planets that are farthest from the sun. The outer solar system consists of Jupiter, Saturn, Uranus, and Neptune, the gas giants. Pluto was considered to be the ninth planet until 2006. Astronomers have since classified it, along with other small icy bodies far from the sun, as a minor planet. Pluto is now considered to be part of the *Kuiper Belt,* which is a vast area beyond the orbit of Neptune that includes many comets. The most distant part of the solar system, the Oort cloud, extends halfway to the nearest star and is also home to many comets.

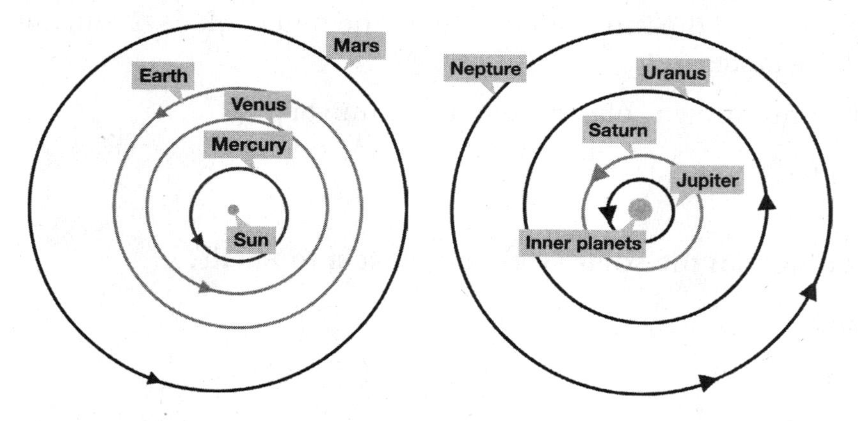

GUIDED PRACTICE

Directions: Using the Standard Review and what you have studied, read each question and circle the letter of the best response.

According to the illustration above, which planet passes closest to Earth?

A Mercury

B Jupiter

C Venus

D Neptune

The correct answer is C. Mercury orbits closer to the sun than Venus (Answer C), so it cannot approach Earth as closely as Venus does. Jupiter (Answer B) and Neptune (Answer D) are part of the outer solar system, which is much more distant from Earth than the inner solar system planets.

Grade 6 – Earth and Space Science: Standard 6 – The Universe

STANDARD PRACTICE

1. Which of these is generally most distant from Earth?

A the outer planets

B the moon

C comets

D the sun

2. Our solar system

F consists of the sun and nine planets.

G includes the planets, but not the sun.

H is made up of the inner solar system, the outer solar system, and other orbiting systems.

J is the only known solar system in the universe.

3. What object in the solar system is closest to Earth?

A Venus

B the sun

C Mars

D the moon

4. a. What objects make up the most distant part of the solar system?

b. What property determines whether an object is part of the solar system or outside the solar system?

Grade 6 – Earth and Space Science: Standard 6 – The Universe

GLE 0607.6.3 Explain how the positional relationships among the earth, moon, and sun control the length of the day, lunar cycle, and year.

STANDARD REVIEW

Our solar system includes the sun, the planets, and many smaller objects. Observations provide evidence that the planets travel around the sun along a path called an *orbit.* One complete trip along an orbit is called a *revolution.* Each planet takes a different amount of time to orbit the sun. Earth's period of revolution is about 365.25 days (one year).

In addition to orbiting the sun, each planet in the solar system also spins on its axis. The spinning of a planet or other body on its axis is called *rotation.* Earth requires 24 hours (one day) to complete its rotation. As it rotates, only one half of the planet faces the sun at a time. The half facing the sun is lit (daytime). The half that faces away from the sun is dark (nighttime).

The moon revolves around Earth approximately every 29 days. This period is called *one month.* As the moon revolves around Earth, the amount of sunlight on the side of the moon that faces Earth changes. These different appearances of the moon result from its changing position relative to Earth and the sun. Every 29 days, the moon's Earthward face changes from a fully lit circle to a thin crescent and then back to a circle. However, because the moon's period of rotation is exactly equal to its period of revolution, we always see the same part of the moon's surface from Earth.

GUIDED PRACTICE

Directions: Using the Standard Review and what you have studied, read each question and circle the letter of the best response.

On a planet with a slower period of rotation than Earth, which of the following would always be true?

- **A** A year would be longer than 365 days.
- **B** A day would longer than 24 hours.
- **C** A day would be shorter than 24 hours.
- **D** A year would be shorter than 365 days.

The correct answer is B. If the planet's rotation is slower than that of Earth, its day would have to be longer than that of Earth, so Answer C is incorrect. Because the period of rotation has no effect on the length of a year, Answers A and D are incorrect.

STANDARD PRACTICE

1. Earth's period of revolution is

A 365.25 days.

B 24 hours.

C 29 days.

D 12 hours.

2. Earth completes one full rotation on its axis every 24 hours. During this period of time, the region near Earth's equator will experience

F 24 hours of daylight.

G 24 hours of darkness.

H approximately 12 hours of daylight and 12 hours of darkness.

J a lunar eclipse.

3. What happens in one year?

A The moon completes one orbit around Earth.

B The sun travels once around Earth.

C Earth revolves once on its axis.

D Earth completes one orbit around the sun.

4. a. What is the difference between a planet's period of rotation and period of revolution?

b. What is the length of Earth's period of rotation? What is the length of its period of revolution?

Grade 6 – Earth and Space Science: Standard 6 – The Universe

GLE 0607.6.4 Describe the different stages in the lunar cycle.

STANDARD REVIEW

From Earth, one of the most noticeable aspects of the moon is its continually changing appearance. Within a month, the moon's Earthward face changes from a fully lit circle to a thin crescent and then back to a circle. These different appearances of the moon result from its changing position relative to Earth and the sun. As the moon revolves around Earth, the amount of sunlight on the side of the moon that faces Earth changes.

The different appearances of the moon, due to its changing position, are called phases. The phases of the moon are shown in the illustration below. When the moon is *waxing,* the sunlit fraction that we can see from Earth is getting larger. When the moon is *waning,* the sunlit fraction is getting smaller. Notice in the figure below that even as the phases of the moon change, the total amount of sunlight that the moon gets remains the same. Half the moon is always in sunlight, just as half of Earth is always in sunlight. However, because the moon's period of rotation is the same as its period of revolution, on Earth you always see the same side of the moon. If you lived on the far side of the moon, you would see the sun for half of each lunar day, but you would never see Earth!

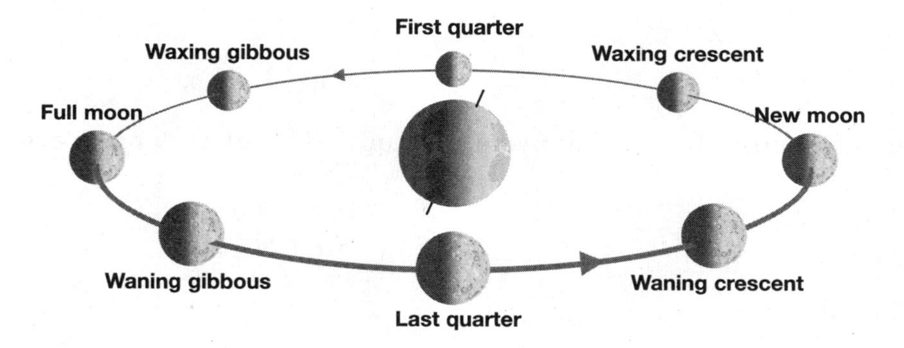

GUIDED PRACTICE

While the moon is waxing, the sunlit part we see

A increases.

B decreases.

C stays the same.

D disappears.

The correct answer is A. A waxing moon is defined as the period when the sunlit portion we see is increasing. Answer B describes a waning moon, not a waxing moon. Answer C is incorrect because the sunlit portion we see is constantly changing, and answer D is incorrect because parts of the moon do not disappear. We cannot see them because there is no light reflecting from them.

Grade 6 – Earth and Space Science: Standard 6 – The Universe

STANDARD PRACTICE

1. Organize these phases of the moon in order, beginning with a full moon.

A full, waning gibbous, first quarter, waxing crescent, new

B full, waning gibbous, new, waxing crescent, first quarter

C full, waxing crescent, new, waning gibbous, first quarter

D full, waxing crescent, waning gibbous, new, first quarter

2. Jared is planning an experiment to track the phases of the moon. Which of the following timekeeping devices would be Jared's best choice?

F stopwatch

G sundial

H calendar

J wristwatch

3. The moon completes a full cycle through all its phases once every

A 24 hours.

B 29 days.

C 16 days.

D 365 days.

4. a. What causes the phases of Earth's moon?

b. Is the far side of the moon always dark? Explain your answer.

Grade 6 – Earth and Space Science: Standard 6 – The Universe

GLE 0607.6.5 Produce a model to demonstrate how the moon produces tides.

STANDARD REVIEW

Tides are caused by the effect of the moon's gravity on Earth. Because water is fluid, it shows the effects of gravity more than the solid parts of the planet do. How often tides occur and the difference in tidal levels depend on the position of the moon as it revolves around Earth. The moon's pull is strongest on the part of Earth directly facing the moon. When part of the ocean is directly facing the moon, the water there bulges toward the moon. These bulges are called *high tides.* It is interesting that there is also a high tide on the opposite side of Earth. This is because the moon does not revolve around the center of Earth. Both bodies actually revolve around the center of their combined masses. This is a point just near Earth's surface. You can model the effect as Earth's gravity pulling one way around this point while the moon pulls the other way. There are two high tides because water is pulled in one direction by the moon and in the opposite direction by Earth.

The sun also affects tides. The sun is much larger than the moon, but much farther away, so the sun's influence on tides is less than the moon's influence. As Earth rotates, the bulge of water moves relative to the surface. That is why there are two tides each day, one for each bulge. The time of the tides changes from day to day as the moon's position changes relative to Earth.

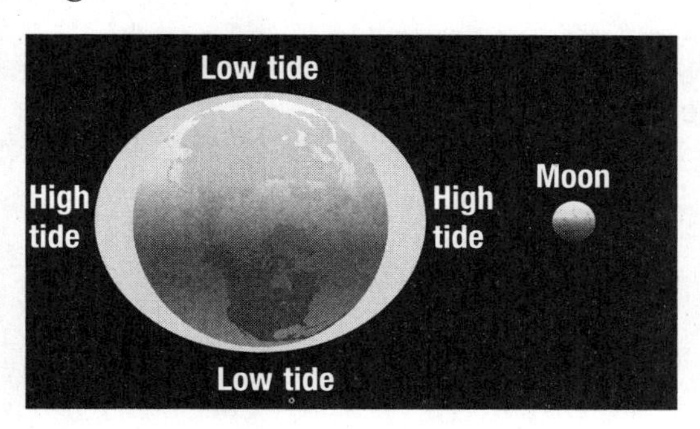

GUIDED PRACTICE

Directions: Using the Standard Review and what you have studied, read each question and circle the letter of the best response.

The moon affects tides more than the sun because the moon

A is closer.

B is made of rock.

C is larger.

D has no atmosphere.

The correct answer is A. Although they appear to be the same size, the sun is much larger than the moon (Answer C) but much farther away. Although Answer B and D correctly describe the moon, these features do not affect tides.

STANDARD PRACTICE

1. What determines tides in Earth's oceans?

A rotation of Earth and the sun's gravitational pull on Earth

B rotation of Earth and the moon's revolution around Earth

C revolution of Earth and the moon around the sun

D rotation of the moon

2. Which of the following is the best description of neap tides?

F tides with the largest daily tidal range

G tides that occur when the sun, Earth, and the moon are aligned

H tides with the smallest daily tidal range

J bulges caused by Earth's rotation and the moon's gravity

3. Tides are at their highest during

A spring tide.

B neap tide.

C a tidal bore.

D the daytime.

4. a. What causes tides and tidal cycles on Earth?

b. Compare and contrast the influences of the sun, the moon, and Earth's rotation on the tides.

Grade 6 – Earth and Space Science: Standard 6 – The Universe

GLE 0607.6.6 Illustrate the relationship between the seasons and the earth-sun system.

STANDARD REVIEW

Energy from the sun heats Earth. The amount of direct solar energy a particular area receives is determined by *latitude*, which measures the distance from the equator. The curve of Earth affects the amount of direct solar energy at different latitudes. Notice that the sun's rays hit the equator directly, at almost a 90° angle. At this angle, a small area of Earth's surface receives more direct solar energy than at a lesser angle. As a result, that area has high temperatures. However, the sun's rays strike the poles at a lesser angle than they do the equator. At this angle, the same amount of direct solar energy that hits the area at the equator is spread over a larger area at the poles. The result is lower temperatures at the poles.

In most places in the United States, the year consists of four seasons. But there are places in the world that do not have such seasonal changes. For example, areas near the equator have approximately the same temperatures and same amount of daylight year-round. Seasons happen because Earth is tilted on its axis at a 23.5° angle. This tilt affects how much solar energy an area receives as Earth moves around the sun. Latitude and the tilt of Earth determine the seasons and the length of the day in a particular area.

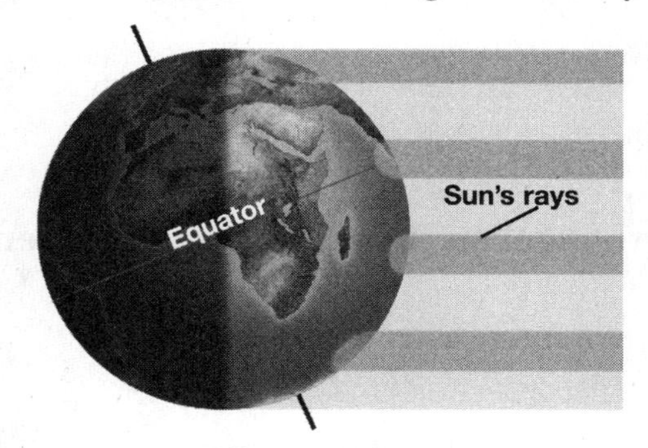

GUIDED PRACTICE

Directions: Using the Standard Review and what you have studied, read each question and circle the letter of the best response.

Which of the following latitudes receive the <u>most</u> direct solar energy?

- **A** latitudes near the north pole
- **B** latitudes near the south pole
- **C** latitudes near mountain ranges
- **D** latitudes near the equator

The correct answer is D. Latitudes near the equator receive direct solar energy during the entire year, while latitudes near the poles (Answers A and B) receive direct sunlight only during half the year. Mountains (Answer C) occur at all latitudes of Earth.

STANDARD PRACTICE

1. When it is mid-winter at the north pole, what is the season in the United States?

A spring

B summer

C autumn

D winter

2. In which of the following is the tilt of Earth's axis considered to have an effect on climate?

F global warming

G the sun's cycle

H the Milankovitch theory

J asteroid impact

3. What part of Earth receives the most direct solar energy?

A the poles

B the equator

C midway between the poles and the equator

D equal at all points on Earth's surface

4. Seasons on Earth are related to the tilt of its axis.

a. Explain why there is a greater seasonal difference in climate between areas at 0° latitude and areas at 45° latitude.

b. How would seasons be different if Earth did not tilt on its axis?

Grade 6 – Earth and Space Science: Standard 6 – The Universe

GLE 0607.6.7 Describe the causes of lunar and solar eclipses.

STANDARD REVIEW

When the shadow of one celestial body falls on another, an *eclipse* occurs. A *solar eclipse* happens when the moon comes between Earth and the sun, and the shadow of the moon falls on part of Earth. Because the moon's orbit is elliptical, the distance between the moon and Earth changes. During an *annular eclipse,* the moon is farther from Earth, and the disk of the moon does not completely cover the disk of the sun. When the moon is closer to Earth, the moon appears to be the same size as the sun. During a *total solar eclipse,* the disk of the moon completely covers the disk of the sun.

A *lunar eclipse* happens when Earth comes between the sun and the moon, and the shadow of Earth falls on the moon. Earth's atmosphere acts like a lens and bends some of the sunlight into Earth's shadow. When sunlight hits the particles in the atmosphere, blue light is filtered out. As a result, most of the remaining light that lights the moon is red. You don't see solar and lunar eclipses every month because the moon's orbit around Earth is tilted—by about 5°—relative to the orbit of Earth around the sun. During most months, the three bodies do not line up in a straight line.

GUIDED PRACTICE

Directions: Using the Standard Review and what you have studied, read each question and circle the letter of the best response.

What type of eclipse occurs when Earth is between the sun and the moon?

A solar eclipse

B annular eclipse

C lunar eclipse

D total eclipse

The correct answer is C. A lunar eclipse occurs when the shadow of Earth blocks sunlight so that it does not reach the moon. The other answers all describe solar eclipses, which occur when the moon is between Earth and the sun.

Grade 6 – Earth and Space Science: Standard 6 – The Universe

STANDARD PRACTICE

1. What causes the moon to appear red during a lunar eclipse?

- **A** Earth's atmosphere filters out blue light.
- **B** Earth's atmosphere filters out red light.
- **C** Light reflects from Earth's surface to the moon.
- **D** The moon blocks blue light from the sun.

2. What arrangement of bodies causes an annular eclipse?

- **F** Earth is between the sun and the moon, and the moon is in a close part of its orbit around Earth.
- **G** Earth is between the sun and the moon, and the moon is in a distant part of its orbit around Earth.
- **H** The moon is between the sun and Earth, and the moon is in a close part of its orbit around Earth.
- **J** The moon is between the sun and Earth, and the moon is in a distant part of its orbit around Earth.

3. For a solar eclipse to occur, which of the following alignments is necessary?

- **A** The moon is located along a straight line between the sun and Earth.
- **B** The Earth is located along a straight line between the sun and the moon.
- **C** The moon is located 90° from a line between the sun and Earth.
- **D** The Earth is located 90° from a line between the sun and the moon.

4. The illustration below shows an eclipse.

a. What type of eclipse is shown in the illustration?

b. Can this eclipse be seen from everywhere on the daylight side of Earth? Explain your answer.

Grade 6 – Earth and Space Science: Standard 8 – The Atmosphere

GLE 0607.8.1 Design and conduct an investigation to determine how the sun drives atmospheric convection.

STANDARD REVIEW

Large masses of air move through the atmosphere by convection when there are differences in pressure. Uneven heating of Earth by the sun's energy creates differences in air pressure. The equator receives more direct solar energy than other latitudes, so air at the equator is warmer and less dense than the surrounding air. This warm, less dense air rises and flows toward the poles, cooling as it moves away from the equator. Air cools as it moves from the equator. When it is colder and denser than the surrounding air, it sinks because of its greater density. The warm air that is pushed upward continues toward the poles. Cold polar air then flows back toward the equator. These large convection currents affect climate around the planet.

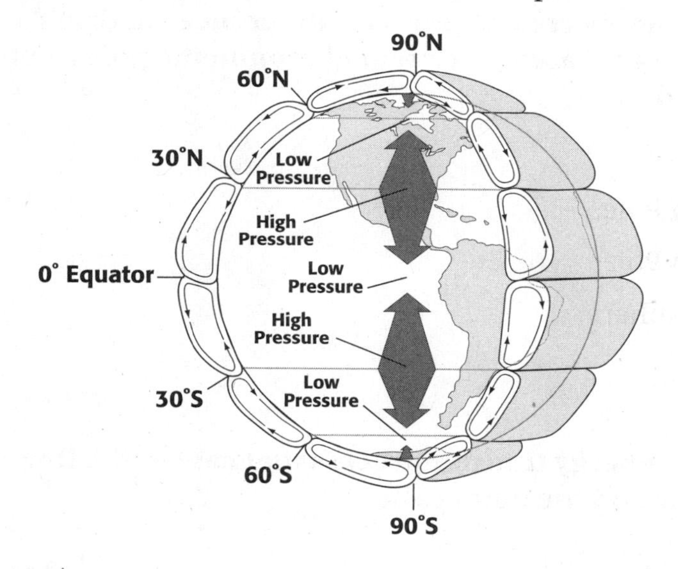

GUIDED PRACTICE

Directions: Using the Standard Review and what you have studied, read each question and circle the letter of the best response.

Warm, less dense air at the equator

- **A** cools and creates an area of low pressure.
- **B** cools and creates an area of high pressure.
- **C** cools and then flows toward the poles.
- **D** rises and flows toward the poles.

The correct answer is D. Because warm air is less dense, it rises before moving away from the equator. Air forms a high pressure area when it cools, making Answer A incorrect, but the warm air does not cool until it moves away from the equator (Answers B and C).

Grade 6 – Earth and Space Science: Standard 8 – The Atmosphere

STANDARD PRACTICE

1. Air is warmer and less dense than surrounding air at the equator because the equator receives more

A wind.

B air pressure.

C solar energy.

D greenhouse gases.

2. The sun's energy creates pressure differences in Earth's atmosphere. As high pressure areas are created around the poles, cold polar air will flow toward

F the equator.

G the North Pole.

H the South Pole.

J the atmosphere.

3. What is the energy transfer process that causes the flow of air between the equator and the poles called?

A conduction

B convection

C radiation

D insulation

4. a. How would winds be affected if Earth's surface were the same temperature everywhere? Explain.

b. What would happen if the poles received more solar energy than was received at the equator?

Grade 6 – Earth and Space Science: Standard 8 – The Atmosphere

GLE 0607.8.2 Describe how the sun's energy produces the wind.

STANDARD REVIEW

Uneven heating of Earth by the sun's energy creates differences in air pressure that power the flow of air between the equator and the poles. Smaller convection currents are also caused by differences in heating in the atmosphere that create areas of high and low pressure. Areas that have lower pressure than their surroundings are called *cyclones.* Cyclones are areas where air masses come together, or *converge,* and rise. Areas that have high pressure are called *anticyclones.* Anticyclones are areas where air moves apart, or *diverges,* and sinks. As the air in the center of a cyclone rises, it cools and forms clouds and rain. The rising air in a cyclone causes stormy weather. In an anticyclone, the air sinks. As the air sinks, it gets warmer and absorbs moisture. The sinking air in an anticyclone brings dry, clear weather.

Thunderstorms are small, intense weather systems that produce strong winds, heavy rain, lightning, and thunder. Thunderstorms can occur along cold fronts, but thunderstorms can develop in other places, where there is warm, moist air near Earth's surface and an unstable atmosphere. The atmosphere is unstable when the surrounding air is colder than the rising air mass. The air mass will continue to rise as long as the surrounding air is colder than the air mass.

GUIDED PRACTICE

Directions: Using the Standard Review and what you have studied, read each question and circle the letter of the best response.

What happens to warm air that enters a cyclone?

A It becomes cooler and sinks.

B It stops moving and forms a stationary air mass.

C It rises and forms clouds.

D It rises and becomes warmer.

The correct answer is C. Warm air rises because it is less dense. As it rises, the air cools, and clouds form. The air does not cool until after it rises (Answers A and D). Because different air masses at different pressures come together in a cyclone, the air is in constant motion (Answer B).

Grade 6 – Earth and Space Science: Standard 8 – The Atmosphere

STANDARD PRACTICE

1. What is the initial energy source for ocean currents and wind?

- **A** air pressure
- **B** the moon's gravity
- **C** the sun
- **D** diverging continental plates

2. Which of these best describes wind?

- **F** moving air
- **G** essential to the water cycle
- **H** static air near the ocean's surface
- **J** a difference in air temperature

3. A severe storm that forms as a rapidly rotating funnel cloud is called a

- **A** hurricane.
- **B** tornado.
- **C** typhoon.
- **D** thunderstorm.

4. a. How does a hurricane form?

b. What happens to a hurricane when it moves over cold water?

Grade 6 – Earth and Space Science: Standard 8 – The Atmosphere

GLE 0607.8.3 Investigate the relationship between currents and oceanic temperature differences.

STANDARD REVIEW

Ocean water contains stream-like movements of water called ocean *currents.* Currents are influenced by a number of factors, including weather, Earth's rotation, and the position of the continents. Surface currents are controlled by three factors: global winds, the Coriolis effect, and continental deflections. These three factors keep surface currents flowing in distinct patterns around Earth. Unlike surface currents, deep currents are not directly controlled by wind. Instead, deep currents form in parts of the ocean where water density increases. Both decreasing the temperature of ocean water and increasing the water's salinity increase the water's density.

Oceans can influence an area's climate. Water absorbs and releases thermal energy much more slowly than dry land does. Because of this quality, water helps to moderate the temperatures of the land around it. Therefore, sudden or extreme temperature changes rarely take place on land near oceans. As surface currents move, they carry warm or cool water to different locations. The surface temperature of the water affects the temperature of the air above it. Warm currents heat the surrounding air and cause warmer temperatures. Cool currents cool the surrounding air and cause cooler temperatures.

GUIDED PRACTICE

Directions: Using the Standard Review and what you have studied, read each question and circle the letter of the best response.

Although the British Isles are at a high latitude, the climate in the Isles is very mild. What is the most likely reason for this?

A The British Isles receive a great deal of sunshine.

B The British Isles are located near a warm-water surface current.

C The British Isles experience a very short winter.

D The British Isles are located near a cold-water surface current.

The correct answer is B. A warm surface current heats the air above it, creating milder climate conditions. A cold current (Answer D) would have the opposite effect. The amount of sunshine (Answer A) cannot be correct because many areas at the same latitude have harsh winters. The length of winter (Answer C) is the same for all areas at the same distance from the equator.

Grade 6 – Earth and Space Science: Standard 8 – The Atmosphere

STANDARD PRACTICE

1. The initial source of energy for ocean currents is

A the rotation of Earth.

B the moon's gravity.

C the sun.

D continental deflections.

2. Southern California has cooler temperatures than Arizona, which is at the same latitude, but farther inland. What causes this temperature difference?

F a cold-water surface current off California's coast

G winds formed high in California's mountains

H the shade provided by the large number of trees in California

J the large longshore current experienced by California

3. How does the surface temperature of the ocean influence the atmosphere?

A As the ocean's surface cools, gases in the atmosphere heat it.

B As the ocean's surface cools, less atmospheric gas is absorbed by the water.

C As the ocean's surface is heated, more water evaporates and enters the atmosphere.

D As the ocean's surface is heated, the air in the atmosphere cools.

4. a. How do oceans receive thermal energy, and how is this energy transferred by ocean currents?

b. What is the name of this type of heat transfer?

Grade 6 – Earth and Space Science: Standard 8 – The Atmosphere

GLE 0607.8.4 Analyze meteorological data to predict weather conditions.

STANDARD REVIEW

To forecast the weather accurately, meteorologists need to measure various atmospheric conditions, such as air pressure, humidity, precipitation, temperature, wind speed, and wind direction. Meteorologists use special instruments to collect data on weather conditions both near and far above Earth's surface.

A tool used to measure air temperature is called a *thermometer.* A *barometer* is an instrument used to measure air pressure. Wind direction can be measured by using a *windsock* or a *wind vane.* A wind vane is shaped like an arrow with a large tail and is attached to a pole. As the wind pushes the tail of the wind vane, the wind vane spins on the pole until the arrow points into the wind. An instrument used to measure wind speed is called an *anemometer,* which consists of three or four cups connected by spokes to a pole. The wind pushes on the hollow sides of the cups and causes the cups to rotate on the pole. The motion sends a weak electric current that is measured and displayed on a dial. *Radar* is used to find the location, movement, and amount of precipitation. It can also detect what form of precipitation a weather system is carrying.

Weather information from all of these instruments is used to make a *weather map.* The map combines the observations from many weather stations around the world as well as observations from satellites. Weather maps usually show lines called *isobars* that connect points of equal air pressure. *Fronts,* where different air masses meet, are also labeled on weather maps. Meteorologists use this information to predict the movement of weather systems.

GUIDED PRACTICE

One tool that meteorologists use to determine the location of a heavy downpour is

A a thermometer.

B an anemometer.

C radar.

D a weather map.

The correct answer is C. Radar readings can show the location of precipitation and the direction in which it is moving. A thermometer (Answer A) measures temperature and an anemometer (Answer B) measures wind speed. These measurements help to predict weather changes but not to find the location of precipitation. The weather map (Answer D) uses the information from radar and other instruments to display weather information but not to find it.

Grade 6 – Earth and Space Science: Standard 8 – The Atmosphere

STANDARD PRACTICE

1. **Which of these instruments measures air pressure?**

A barometer
B anemometer
C Doppler radar
D wind sock

Use this weather map to answer questions 2 and 3.

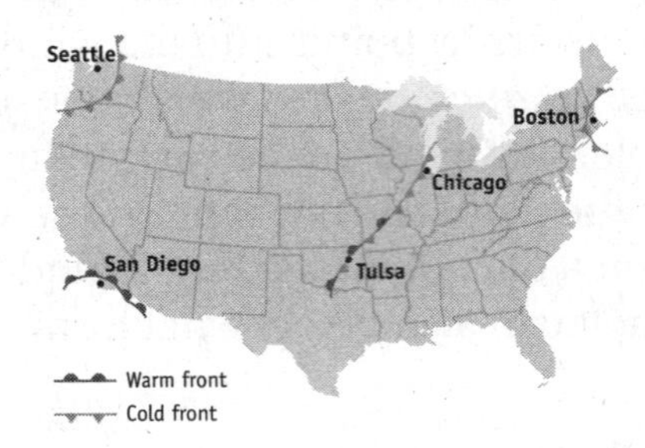

2. **Which city is most likely to have thunderstorms approaching?**

F Boston
G San Diego
H Seattle
J Chicago

3. **What type of front has formed near Tulsa?**

A warm front
B cold front
C stationary front
D alternating front

4. **a. Why do meteorologists use information from many weather stations to make a local prediction?**

b. How is information about the movement of warm and cold fronts useful for weather prediction?

GLE 0607.10.1 Compare and contrast the three forms of potential energy.

STANDARD REVIEW

Kinetic energy is the energy of motion, but not all energy has to do with motion. *Potential energy* is the energy an object has because of its position. There are three forms of potential energy. When you lift an object, you do work on it. You use a force that opposes the force of gravity. You transfer energy to the object and give the object *gravitational potential energy.* The amount of gravitational potential energy that an object has depends on its weight and its height. A rubber band can be used to show another example of potential energy. When the rubber is stretched, work is done on it, so energy is transferred to the rubber band. This type of potential energy is called *elastic potential energy.* When the rubber band is released, the stored energy becomes *kinetic energy.* The third type of potential energy exists in chemical compounds. When chemical bonds form between atoms, work is done to join them together. *Chemical energy* is the energy of a compound that changes as its atoms are rearranged. Chemical energy is a form of potential energy because it depends on the position and arrangement of the atoms in a compound.

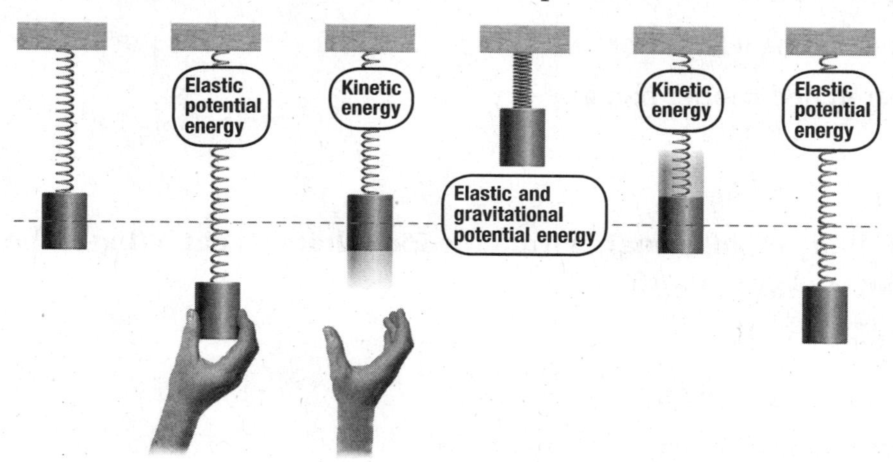

GUIDED PRACTICE

Directions: Using the Standard Review and what you have studied, read each question and circle the letter of the best response.

Gravitational potential energy depends on

A mass and velocity. **B** mass and weight.

C weight and height. **D** height and distance.

The correct answer is C. Gravitational potential energy is a function of weight (mass times the acceleration of gravity) and height. Answer A is not correct because velocity refers to motion and kinetic energy. Answer B does not include height, and answer D does not include weight, both of which must be known to calculate gravitational potential energy.

STANDARD PRACTICE

1. Sam studied kinetic and potential energy by observing apples on a tree during a field investigation. What type of energy do these apples have?

A kinetic energy

B mechanical energy

C gravitational potential energy

D elastic potential energy

2. Which of the following is <u>not</u> an example of potential energy?

F a ball traveling at 10 m/s

G a compressed spring

H stress along a fault line

J a stretched rubber band

3. Your body obtains energy from the food that you eat. What is the form of the energy in food?

A kinetic energy

B elastic potential energy

C gravitational potential energy

D chemical energy

4. a. Photosynthesis uses sunlight when plants make glucose from carbon dioxide and water. What form of energy does the glucose have?

b. Describe why chemical energy is a form of potential energy.

Grade 6 – Physical Science:
Standard 10 – Energy

GLE 0607.10.2 Analyze various types of energy transformations.

STANDARD REVIEW

There are many forms of energy, and during the transfer of energy it can be converted from one form to another.

Mechanical energy is the total energy of motion and position of an object. Both potential energy and kinetic energy are kinds of mechanical energy. Mechanical energy can be all potential energy, all kinetic energy, or some of each. *Thermal energy* is all of the kinetic energy due to the random motion of the particles that make up an object. *Chemical energy* is the energy of a compound that changes as its atoms are rearranged. Chemical energy is a form of potential energy because it depends on the position and arrangement of the atoms in a compound. *Electrical energy* is the energy of moving electrons. *Sound energy* is caused by an object's vibrations. When you stretch a guitar string, the string stores potential energy. When you let the string go, this potential energy is turned into kinetic energy, which makes the string vibrate. The string also transmits some of this kinetic energy to the air around it. The air particles also vibrate and transmit this energy to your ear. *Light energy* is produced by the vibrations of electrically charged particles. Like sound vibrations, light vibrations cause energy to be transmitted.

Energy can be changed from one form to another. For example, when you light a match, chemical energy of the compounds on the head of the match and in the wood is changed to thermal energy, the energy of motion of particles. In a hydroelectric power plant, the gravitational potential energy of water behind a dam is converted into kinetic energy, which is then converted into electrical energy by the generators.

Table 1 Some Conversions of Electrical Energy	
Alarm clock	electrical energy → light energy and sound energy
Battery	chemical energy → electrical energy
Light bulb	electrical energy → light energy and thermal energy
Blender	electrical energy → kinetic energy and sound energy

GUIDED PRACTICE

Directions: Using the Standard Review and what you have studied, read each question and circle the letter of the best response.

Which energy transformation occurs when a television speaker produces music?

A electrical energy → sound energy

B elastic potential energy → sound energy

C thermal energy → sound energy

D sound energy → electrical energy

The correct answer is A. Electrical energy enters the speaker and is converted to kinetic energy of moving speaker parts, which is then converted to sound energy. The speaker does not store energy by compression or stretching, so Answer B is incorrect. The energy entering the speaker is not thermal energy (Answer C) or sound energy (Answer D).

Grade 6 – Physical Science: Standard 10 – Energy

STANDARD PRACTICE

1. What determines an object's thermal energy?

A the motion of its particles

B its size

C its potential energy

D its mechanical energy

2. What energy transfer happens when you plug in a blender?

F Electrical energy becomes light energy and sound energy.

G Electrical energy becomes kinetic energy and sound energy.

H Electrical energy becomes light energy and thermal energy.

J Chemical energy becomes electrical energy.

3. The energy associated with the movement of an object is

A potential energy.

B kinetic energy.

C nuclear energy.

D magnetic energy.

4. a. How does electrical energy get to your home?

b. What happens when you use electrical energy?

Grade 6 – Physical Science: Standard 10 – Energy

GLE 0607.10.3 Explain the principles underlying the Law of Conservation of Energy.

STANDARD REVIEW

An energy conversion is a change from one form of energy to another. Any form of energy can change into any other form of energy. For example, kinetic energy can change into potential energy, and vice versa. Chemical energy in food can be converted into energy that your body can use, and chemical energy in fuels can be converted into thermal energy by burning the fuels. Plants convert light energy into chemical energy during photosynthesis.

A *closed system* is a group of objects that transfer energy only to each other. For example, a closed system that involves a roller coaster consists of the track, the cars, and the air around them. On a roller coaster, some mechanical energy (the sum of kinetic and potential energy) is always converted into thermal energy because of friction. Sound energy also comes from the energy conversions in a roller coaster. If you add together the cars' kinetic energy at the bottom of the first hill, the thermal energy due to overcoming friction, and the sound energy made, you end up with the same total amount of energy as the original amount of potential energy. In other words, energy is conserved and not lost.

According to the *Law of Conservation of Energy,* energy cannot be created or destroyed. The total amount of energy in a closed system is always the same. Energy can change from one form to another, but all of the different forms of energy in a system always add up to the same total amount of energy, no matter how many energy conversions take place.

GUIDED PRACTICE

Directions: Using the Standard Review and what you have studied, read each question and circle the letter of the best response.

During energy changes, part of the energy is always converted to

A nuclear energy.

B potential energy.

C thermal energy.

D mechanical energy.

The correct answer is C. Whenever energy is converted, part of the change becomes thermal energy as a result of friction. Nuclear energy (Answer A) is only involved in transformations of the nucleus of an atom. Potential and mechanical energy (Answers B and D) are involved in changes of position or motion but not in all energy changes.

Grade 6 – Physical Science:
Standard 10 – Energy

STANDARD PRACTICE

1. Which of these statements describes the Law of Conservation of Energy?

A No machine is 100% efficient.
B Energy is neither created nor destroyed.
C The energy resources of Earth are limited.
D The energy of a system is always decreasing.

2. Perpetual motion is impossible because

F things tend to slow down.
G energy is lost over time.
H machines are very inefficient.
J friction converts some energy to thermal energy.

3. As a roller coaster moves down a hill, potential energy is converted into kinetic energy, thermal energy, and sound energy. What is true about the total energy of this system?

A Energy is gained as the roller coaster moves.
B Energy is lost as the roller coaster moves.
C The total energy remains the same as the roller coaster moves.
D The total energy constantly changes as the roller coaster moves.

4. Imagine that you drop a ball. It bounces a few times and then it stops. Your friend says that the energy that the ball had is gone.

a. What kind of energy did the ball have, and where did the energy go?

b. Evaluate your friend's statement based on energy conservation.

Grade 6 – Physical Science: Standard 12 – Forces in Nature

GLE 0607.12.1 Describe how simple circuits are associated with the transfer of electrical energy.

STANDARD REVIEW

Just like a roller coaster, an electric circuit always forms a loop—it begins and ends in the same place. Because a circuit forms a loop, a circuit is a closed path. So, an *electric circuit* is a complete, closed path through which electric charges flow.

All circuits need three basic parts: an energy source, wires, and a load. Loads, such as a light bulb or a radio, are connected to the energy source by wires. *Loads* change electrical energy into other forms of energy. These other forms might include thermal energy, light energy, sound energy, or mechanical energy.The loads in a circuit can be connected in different ways. Circuits are often divided into two types: series circuits and parallel circuits. One of the main differences in these circuits is the way in which the loads are connected to one another. A *series circuit* is a circuit in which all parts are connected in a single loop. There is only one path for charges to follow, so the charges moving through a series circuit must flow through each part of the circuit. If the circuit is broken, the current stops. When light bulbs are connected in a series circuit, all of the bulbs go out if one of them stops working.

If all of the lights in your home were connected in series, they would all have to be turned on or all off. Instead of being wired in series, circuits in buildings are wired in parallel. A *parallel circuit* is a circuit in which loads are connected side by side. Charges in a parallel circuit have more than one path on which they can travel. One part of the circuit can be turned off, but the other parts stay on. That means you can turn the overhead light off but keep the television on, even if the current for both the overhead light and the television comes from the same wire in the main circuit box of your home.

GUIDED PRACTICE

Directions: Using the Standard Review and what you have studied, read each question and circle the letter of the best response.

Which part of a circuit changes electrical energy into another form of energy?

A energy source

B wire

C switch

D load

The correct answer is D. The load resists the flow of electric charges and converts their electrical energy into another form of energy, such as sound, heat, or light energy. The source (Answer A) refers to the part of the circuit where energy is converted to electrical energy; the wire (Answer B) carries the electrical energy from the source to the load; and a switch (Answer C) is a device used to break the circuit and stop the flow of energy.

Grade 6 – Physical Science: Standard 12 – Forces in Nature

STANDARD PRACTICE

1. In the figure below, which bulb burning out would mean that no current could flow through the circuit?

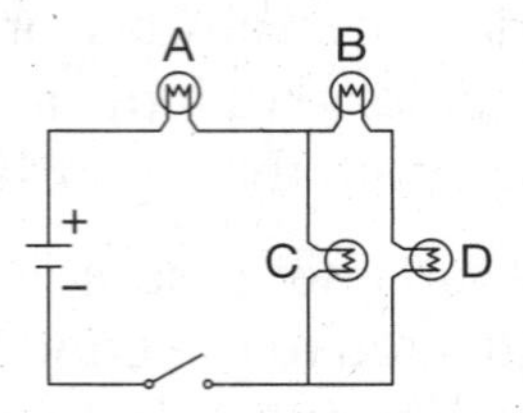

A bulb A

B bulb B

C bulb C

D bulb D

2. If all the electronics in your refrigerator were in a series circuit, what would happen if the light bulb in your refrigerator burned out?

F Another light bulb would automatically replace it.

G The refrigerator would stop working.

H The freezer would get too cold.

J The refrigerator would get too cold.

3. In which circuit below will bulb C remain lit if bulb A burns out?

A

B

C

D

4. a. What is the source of energy in an electric current?

b. Why does an open switch stop the transfer of energy?

Grade 6 – Physical Science: Standard 12 – Forces in Nature

GLE 0607.12.2 Explain how simple electrical circuits can be used to determine which materials conduct electricity.

STANDARD REVIEW

An *electrical conductor* is a material in which charges can move easily. Many metals are good conductors because some of their electrons are free to move. Conductors are used to make wires. For example, a lamp cord has metal wire and metal prongs. Copper, aluminum, and mercury are good conductors. An *electrical insulator* is a material in which charges cannot move easily. Insulators do not conduct charges very well because their electrons cannot flow freely. The electrons are tightly held in the atoms of the insulator. The insulating material in a lamp cord stops charges from leaving the wire and protects you from electric shock. Plastic, rubber, glass, wood, and air are good insulators.

Resistance is one of the factors that determines the amount of current carried in a wire. Resistance is the opposition to the flow of electric charge. You can think of resistance as "electrical friction." Higher resistance of a material causes reduced current in the material. So, if the voltage doesn't change, as resistance goes up, current goes down. An object's resistance depends on its material, thickness, length, and temperature.

Very good conductors, such as copper, have low resistance. Poor conductors, such as iron, have higher resistance. The resistance of insulators is so high that electric charges cannot flow in them at all. Materials with low resistance, such as copper, are used to make wires. Materials with higher resistance are also useful in circuits. For example, the high resistance of the tungsten filament in a light bulb causes the light bulb to heat up and give off light. Other poor conductors convert electrical energy into thermal energy to heat a room or cook food.

GUIDED PRACTICE

Directions: Using the Standard Review and what you have studied, read each question and circle the letter of the best response.

In an electric heater, the coil that glows and gets very hot is made of a material that

A is a good electrical conductor.

B is a poor electrical conductor.

C acts as an insulator.

D carries the current without resistance.

The correct answer is B. The heating element of an electric heater is a poor conductor that becomes hot as it resists the flow of electrons. A good electrical conductor carries the current without resisting it, so it does not become hot (Answers A and D) while an insulator does not allow any current at all and so it also does not become hot (Answer C).

STANDARD PRACTICE

1. What type of material allows electric charges to flow easily?

A conductor

B insulator

C battery

D switch

2. A complete, closed path through which an electric current can flow is a

F load.

G resistor.

H wire.

J circuit.

3. What property do all good conductors of electric current have in common?

A They are metals.

B Charges flow through them without resistance.

C The become very hot when a current passes through them.

D They are coated with insulators.

4. a. What is the purpose of the copper or aluminum wire inside the power cord on an electric appliance?

b. What is the purpose of the plastic coating around the copper or aluminum wire?

TCAP Test Preparation
Practice Test A

1. **The illustration below shows the relative positions of the sun and Earth during different times of the year.**

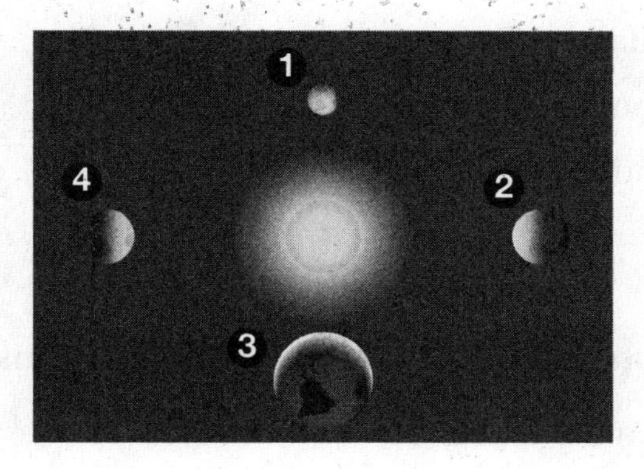

At what position does the South Pole receive almost 24 hours of daylight?

A Position 1

B Position 2

C Position 3

D Position 4

2. **An environmental scientist suspects that acid precipitation is beginning to affect certain lakes in Texas. What is the best way to test this hypothesis?**

F Do library research on the harmful effects of acid precipitation in lakes.

G Experiment with acid precipitation on water plants native to Texas.

H Count the number of water-plant species found in a Texas lake.

J Collect lake water samples, and test the pH of each sample.

3. What are the two main factors of climate?

A air masses and temperature

B wind speed and elevation

C latitude and cloud formation

D temperature and precipitation

4. Which of these statements describes the law of conservation of energy?

F No machine is 100% efficient.

G Energy is neither created nor destroyed.

H The energy resources of Earth are limited.

J The energy of a system is always decreasing.

5. What is the volume of liquid in this graduated cylinder?

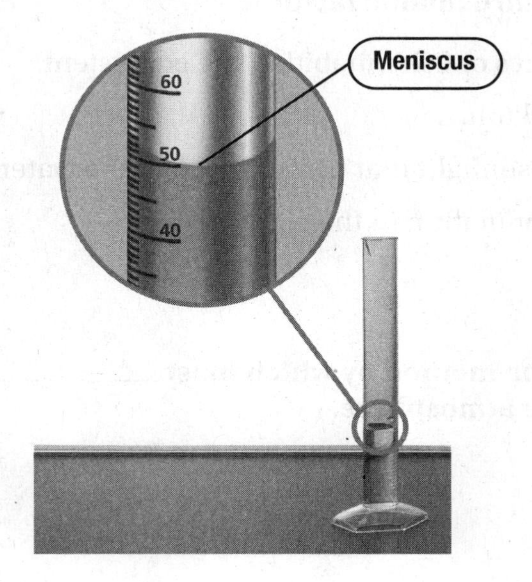

A 49 mL

B 50 mL

C 53 mL

D 55 mL

6. Scientific results must be reproducible, so scientists must keep

F everything they make in a lab.

G all of the books they used for reference.

H photographs of every process.

J accurate records.

7. Water depth and temperature are two abiotic factors in marine ecosystems. A third abiotic factor is

A the many species of fish inhabiting the ecosystem.

B algae and plankton.

C the amount of sunlight that passes through the water.

D the alligator population in the ecosystem.

8. Convection is the method by which most ____________ energy circulates in the atmosphere.

F kinetic

G thermal

H solar

J potential

9. **Imagine that you are an ecologist cataloging the interactions in a salt-marsh community. Look at the illustration of some of the organisms that live in a salt marsh. Which is the abiotic factor in the environment?**

The Salt-Marsh Ecosystem

A cordgrass

B crab

C sparrow

D water

10. **The sun is the initial source of energy for**

F the rotation of Earth.

G the moon's gravity.

H ocean currents.

J continental deflections.

11. Which of the following happens in one year?

A The moon completes one orbit around Earth.

B The sun travels once around Earth.

C Earth revolves once on its axis.

D Earth completes one orbit around the sun.

12. When the Northern Hemisphere is tilted away from the sun,

F Earth is at its farthest location from the sun.

G the equinoxes occur.

H the North Pole experiences its coldest temperatures.

J the Northern Hemisphere experiences the summer solstice.

13. The table below shows a series of mass measurements of a rock by four students. Use an actual mass value of 15.5 grams to answer the question.

Mass of Rock				
	Trial 1	Trial 2	Trial 3	Trial 4
Student A	13.8 g	16.3 g	16.9 g	14.5 g
Student B	15.1 g	15.3 g	15.6 g	15.4 g
Student C	12.8 g	12.4 g	12.6 g	12.5 g
Student D	11.9 g	15.5 g	15.5 g	15.5 g

Which student's measurements could be described as accurate, but not precise?

A Student A

B Student B

C Student C

D Student D

14. As water near the ocean surface cools in the winter, to where is <u>most</u> of its energy transferred?

F nearby land

G the deeper parts of the ocean

H outer space

J the atmosphere

15. The terrestrial and aquatic food webs are connected to one another by which of these processes?

A erosion of rocks and minerals by streams

B leaves and other organic matter that decay in water

C the life cycle of phytoplankton

D benthic organisms

16. How often Earth's tides occur and the difference in tide levels is determined by

F the sun's pull on Earth

G the sun's pull on the moon

H the position of the moon as it revolves around Earth

J the position of Earth as it revolves around the sun

17. **When the moon is waxing, the sunlit part we see**

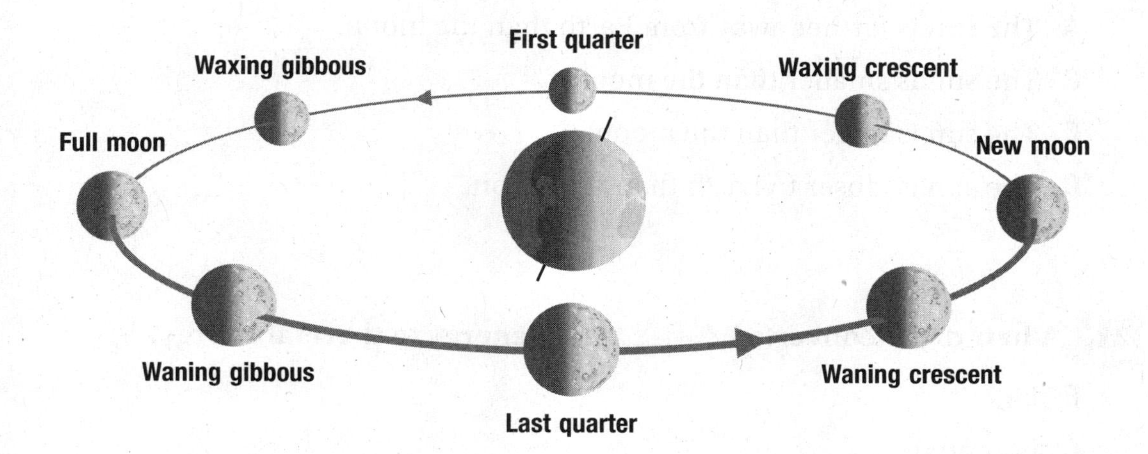

A increases.

B decreases.

C stays the same.

D disappears.

18. **What causes the moon to appear red during a lunar eclipse?**

F Earth's atmosphere filters out blue light.

G Earth's atmosphere filters out red light.

H Light reflects from Earth's surface to the moon.

J The moon blocks blue light from the sun.

19. While the sun has an effect on Earth's tides, why isn't it as strong on the moon's effect?

A The sun is farther away from Earth than the moon.

B The sun is smaller than the moon.

C The sun is larger than the moon.

D The sun is closer to Earth than the moon.

20. A hair dryer converts ____________ energy to thermal energy.

F kinetic

G potential

H electrical

J work

21. Where does energy first come into an ecosystem?

A Producers make energy.

B Scavengers obtain energy from dead plants.

C Consumers obtain energy from animals.

D Sunlight provides energy to producers.

22. Which of the following organisms is a scavenger?

F hawk

G mouse

H grasses

J vulture

23. The diagram below shows a food web for a certain ecosystem. Which of the following best describes what may happen if pollution caused the fish population to decrease?

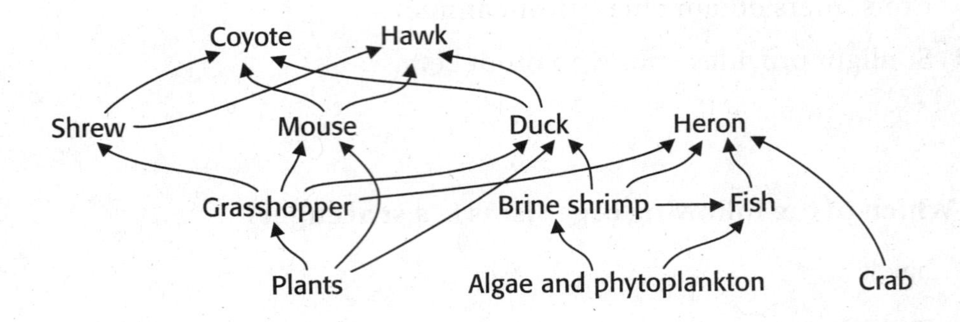

A The brine shrimp population may decrease.

B The mouse population may increase.

C The heron population may decrease.

D The crab population may increase.

24. Nathan sets a pendulum in motion. After several minutes, the pendulum comes to a stop. Which of the following explains why the pendulum stopped?

F All of the pendulum's kinetic energy was converted to potential energy.

G All of the pendulum's energy was lost as thermal energy due to friction.

H All of the pendulum's energy was destroyed by friction.

J All of the pendulum's potential energy was converted to mechanical energy.

25. Chris made the following table as part of a field investigation on the solar system. According to his data, which outer planet has the largest diameter?

Data for the Outer Planets			
Planet	Diameter (km)	Average Distance From the Sun (km)	Period of Revolution (Earth Years)
Jupiter	142,984	778,600,000	12
Saturn	120,536	1,433,500,000	29
Uranus	51,118	2,872,500,000	84
Neptune	49,528	4,495,100,000	164

A Jupiter

B Saturn

C Uranus

D Neptune

26. The length of a day is based on the amount of time that

F Earth takes to orbit the sun one time.

G Earth takes to rotate once on its axis.

H the moon takes to orbit Earth one time.

J the moon takes to rotate once on its axis.

27. Where is the Kuiper Belt located?

A between the orbits of Mars and Jupiter

B around the planet Saturn

C beyond the orbit of Neptune

D near the center of the Milky Way galaxy

28. A scientist finds that the results of his experiment are inconsistent with a long-held theory. What should he do next?

F develop additional experiments to gather more data

G throw out his results and follow the long-held theory

H criticize other scientists for believing the long-held theory

J modify the theory and teach all of his students the revised theory

29. **Where are the biomes containing the largest number of plant and animal species located?**

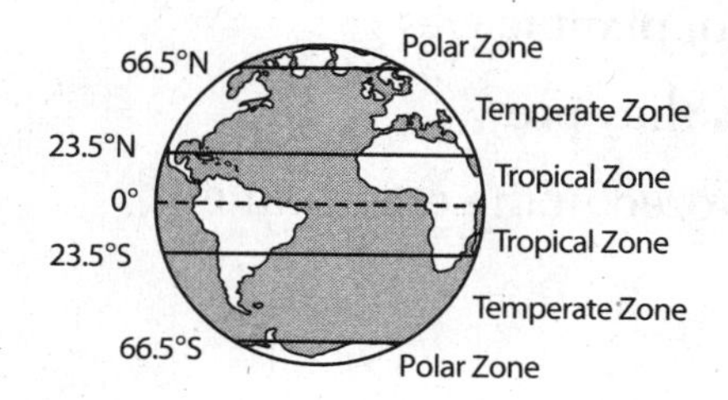

A in the polar zones

B only in the northern portion of the tropical zones

C in the tropical zones

D in the temperate zones

30. **How might a scientist accidentally introduce bias into an experiment?**

F by working with several other scientists instead of working alone

G by writing down only the observations that are most interesting to him

H by having a control group that differs from the experimental groups by only one factor

J by working on a scientific problem at the same time that another research group is working on the same problem

31. What is the most direct source of energy for predators?

A solar energy

B chemical energy of plant leaves

C animals on which they prey

D thermal energy stored in the environment

32. How does scientific writing differ from literary writing?

F The two types of writing have different purposes.

G Scientific reports are always written in English.

H Science writing is shorter than literary writing.

J Science writing is about facts, and literary writing is fiction.

33. **What is the object shown in the illustration below?**

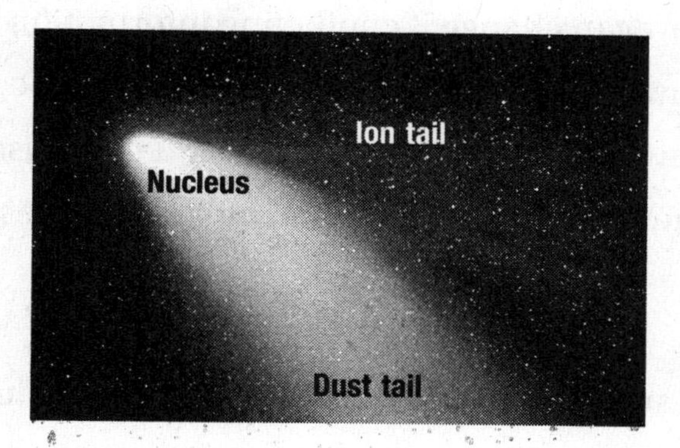

A meteor
B comet
C asteroid
D Oort Cloud

34. **When an archer bends a bow, what type of energy increases in the bow?**

F gravitational potential energy
G elastic potential energy
H kinetic energy
J thermal energy

35. Under what circumstances would a scientific law change?

A A scientific law can change if conflicting information is discovered.

B A scientific law can only change if all scientists agree to change it.

C A scientific law will change whenever new data is discovered.

D A scientific law will change when a majority of scientists agree to change it.

36. How does the surface temperature of the ocean influence the atmosphere?

F As the ocean's surface cools, gases in the atmosphere heat it.

G As the ocean's surface cools, less atmospheric gas is absorbed by the water.

H As the ocean's surface is heated, more water evaporates and enters the atmosphere.

J As the ocean's surface is heated, the air in the atmosphere cools.

37. **This graph shows the volume of water flowing in the Yakima River during a 12-month period. Which conclusion would be supported by the data?**

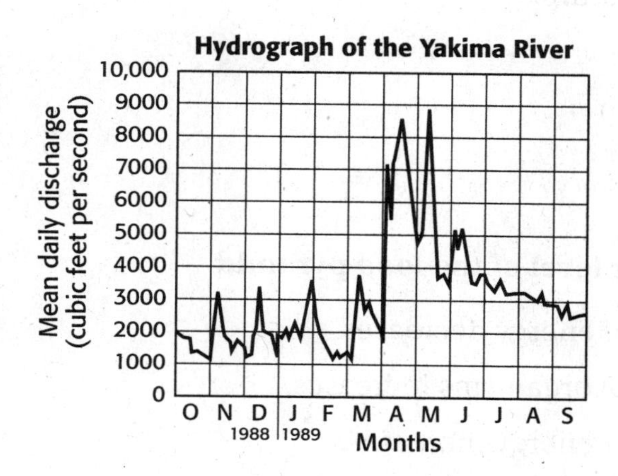

A The water is warmest in late summer.

B The amount of water in the river increases in spring.

C There is more rain in spring.

D The flow decreases in winter because wind patterns change.

38. **Which of the following is a transformation of potential energy to kinetic energy?**

F one end of a lever going up when you push down on the other end

G a sound causing your eardrum to vibrate

H an apple hanging on a branch

J a ball rolling downhill

39. What can you measure using a barometer?

A air temperature

B water temperature

C air pressure

D wind direction

40. At each higher level of the food pyramid

F the amount of energy decreases.

G the number of organisms increases.

H the amount of energy increases.

J the size of organisms increases.

41. **Based on the direction of the prevailing winds and the fact that northern Africa is near the equator, predict the climate of the Sahara.**

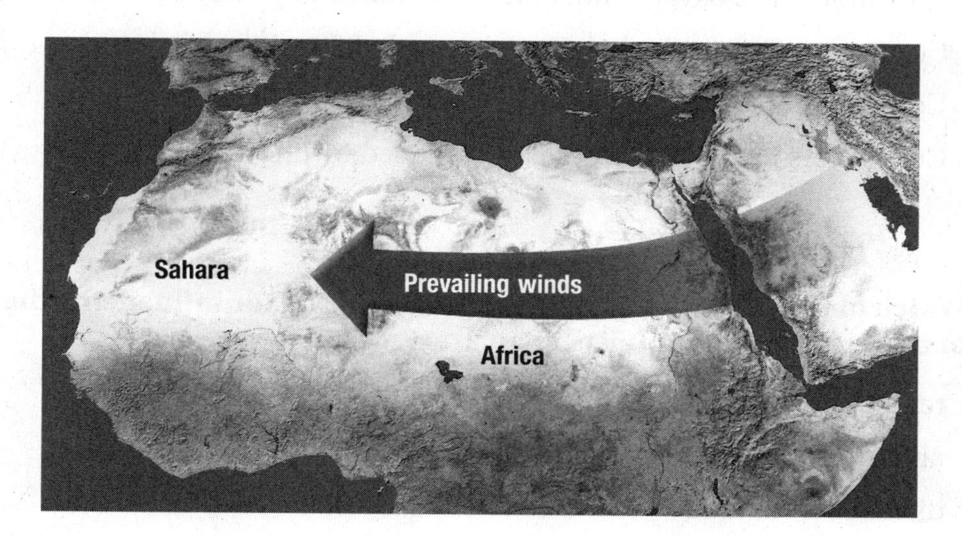

A hot and dry

B cold and dry

C hot and wet

D cold and wet

42. **What are the <u>most</u> important producers in the ocean?**

F phytoplankton

G seaweed

H zooplankton

J sea grasses

43. How is photosynthesis important to consumers?

A Consumers use oxygen and glucose produced by photosynthesis.

B Consumers supply the light energy needed for photosynthesis.

C Consumers supply the water needed for photosynthesis.

D Consumers use the chloroplasts produced by photosynthesis.

44. Which of the following is not a major abiotic factor influencing the kind of marine biome that exists in an area?

F temperature

G amount of sunlight penetrating the water

H distance from land

J type of sand on the ocean bottom

45. Which of the following organisms has the same ecological role as grass on a prairie?

A algae in the ocean

B coyotes in the chaparral

C lions in the savanna

D fungi in a forest

46. The sun provides the __________ that is responsible for all of Earth's weather.

F air pressure

G energy

H gravitational pull

J ocean current

47. **When Earth is oriented, as shown in the illustration below, what is the season in South Africa?**

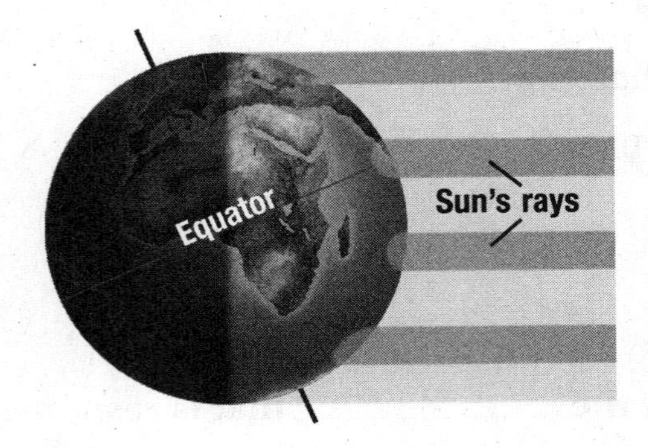

A spring
B summer
C autumn
D winter

48. **Twice a year the length of the day and night are equal. The term for this occurrence is**

F equal distance.
G equal night.
H equinox.
J axis tilt.

49. Which of the following statements describes how the surface temperature of the ocean influences the atmosphere?

A As the ocean's surface is heated, more water evaporates and enters the atmosphere.

B As the ocean's surface cools, gases in the atmosphere heat it.

C As the ocean's surface cools, less atmospheric gas is absorbed by the water.

D As the ocean's surface is heated, the air in the atmosphere cools.

50. Which of the following statements describes observation in science?

F a method scientists use to study organism behavior

G the act of using the senses to gather information

H a way to gather evidence to prove a hypothesis

J all of the above

51. The table below indicates the density compared to mass of Earth (E), Uranus (U), and Neptune (N). The size of the circles represents the diameter of each planet.

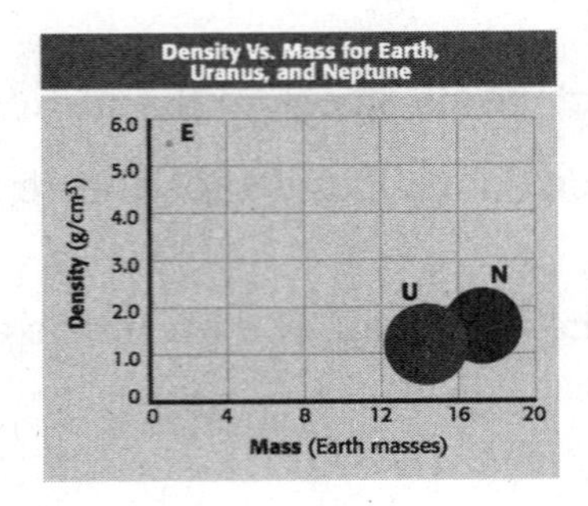

Based on the table, what conclusion can you draw about the planets?

A Earth is larger and denser than either Uranus or Neptune.

B Earth is larger and less dense than either Uranus or Neptune.

C Earth is smaller and denser than either Uranus or Neptune.

D Earth is smaller and less dense than either Uranus or Neptune.

52. Which one of the following <u>cannot</u> be left out of an accurate summary of a scientific investigation?

F supplementary mathematics

G sequence of steps

H historical background

J basic science

53. Temperature is proportional to the average kinetic energy of particles in an object. Thus, an increase in temperature results in a(n)

A increase in mass.

B decrease in average kinetic energy.

C increase in average kinetic energy.

D decrease in mass.

54. Which of these is an example of technology?

F development of the law of Conservation of Energy

G a spacecraft that flies to Neptune to take pictures

H the chemical elements

J insects that live in colonies

55. What causes the constellations that are visible in the night sky to change as the seasons change?

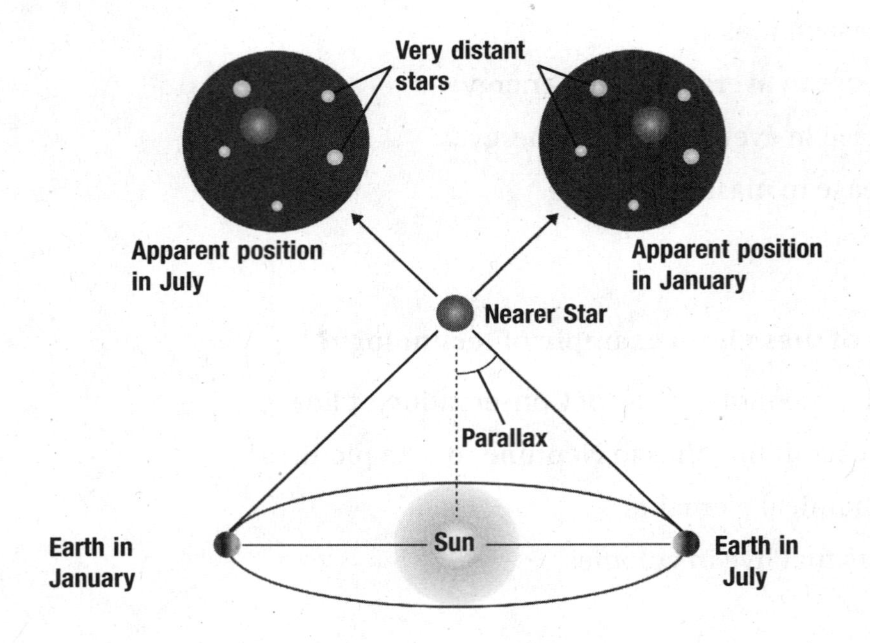

A Earth revolving around the sun

B the light from the stars bending as it travels through Earth's atmosphere

C Earth rotating on its axis

D the expansion of the universe

56. Marla and Ali are going to submit separate investigations to the science fair. They want to avoid bias in the design of their experiments. Which of the following would <u>best</u> help them do that?

F Marla designs both experiments.

G Ali designs both experiments.

H They both choose experiments they have previously performed.

J They help design each other's experiment.

57. Which of these objects is closest to Earth?

A the moon

B the sun

C Venus

D Mars

58. Clouds form in the atmosphere through which process?

F precipitation

G respiration

H condensation

J transpiration

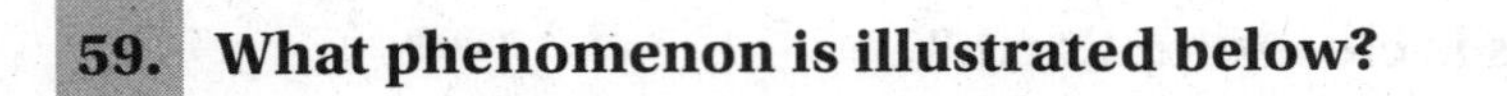

59. What phenomenon is illustrated below?

A total solar eclipse

B annular solar eclipse

C lunar eclipse

D full moon

60. Which of the following is an example of an unintended consequence of technology?

F A new pen is available in only one color of ink.

G A safer car also uses more gasoline.

H A new type of window provides better insulation to a house.

J A change in the design of a generator makes it more efficient.

61. What force prevents the moon from moving away from Earth?

A tides

B electromagnetism

C inertia

D gravity

62. Which of the following is not a form of potential energy?

F gravitational energy

G chemical energy

H thermal energy

J elastic energy

63. At what position is it winter in the Southern Hemisphere?

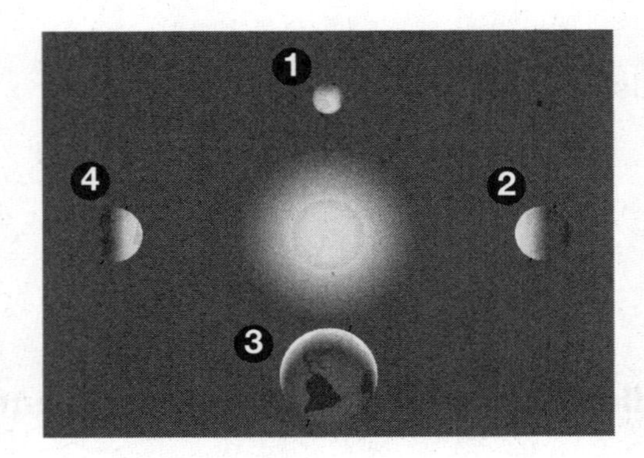

A Position 1

B Position 2

C Position 3

D Position 4

64. Most comets seen from Earth follow an orbit that

F stays between Mars and Jupiter.

G is very close to Earth's orbit.

H is very elliptical.

J is nearly circular.

65. Air at the equator is warmer and less dense than surrounding air because the equator receives more

A wind.

B air pressure.

C solar energy.

D greenhouse gases.

66. What is the source of energy that fuels large storms such as hurricanes?

F evaporation of ocean water

G condensation of water vapor

H solar heating of the atmosphere

J the greenhouse effect

67. What is the maximum wind speed of a tornado that has a rating of F-3?

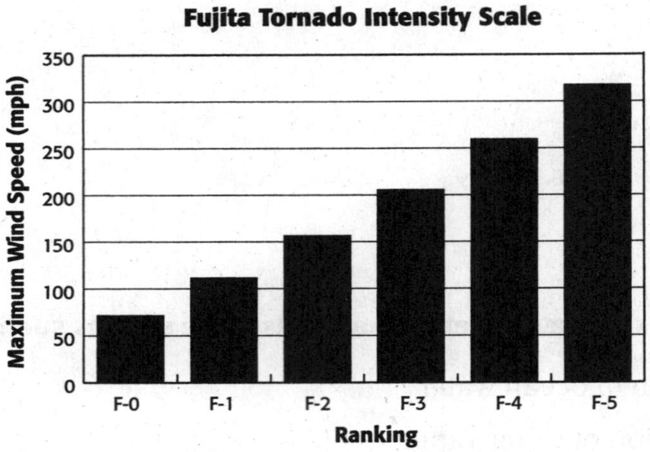

Fujita Tornado Intensity Scale

Maximum Wind Speed (mph): 0, 50, 100, 150, 200, 250, 300, 350

Ranking: F-0, F-1, F-2, F-3, F-4, F-5

A 157 miles per hour

B 206 miles per hour

C 206 kilometers per hour

D 260 kilometers per hour

68. A severe thunderstorm is in the area. A meteorologist says that a cold air mass has moved in. This cold air mass has forced a warm air mass upward. As the warm air cools, it releases energy in the form of severe weather. What is the name for this type of front?

F cold front

G warm front

H occluded front

J stationary front

69. You can tell whether a meter is analog or digital because a digital meter

A plugs into an electric outlet.

B shows readings as numbers.

C can be connected to a computer.

D measures very small amounts.

70. Of these planets, the one that would be classified as a terrestrial planet is

F Mercury.

G Neptune.

H Saturn.

J Uranus.

71. Stars appear to move across the night sky because of

A the rotation of Earth on its axis.

B the movement of the Milky Way galaxy.

C the movement of stars in the universe.

D the revolution of Earth around the sun.

72. Dublin, Ireland is located at a higher latitude than New York City, but Dublin has a warmer average winter temperature. What accounts for the difference in average temperatures between the two cities?

F Dublin's climate is affected by the warm waters of the Gulf Stream.

G Island cities have milder winters than the mainland cities.

H The cities are at different elevations.

J Dublin gets more hours of sunlight in winter than New York City.

73. **In the figure below, warm water expands and rises above cool water. Which physical property describes this relationship between mass and volume?**

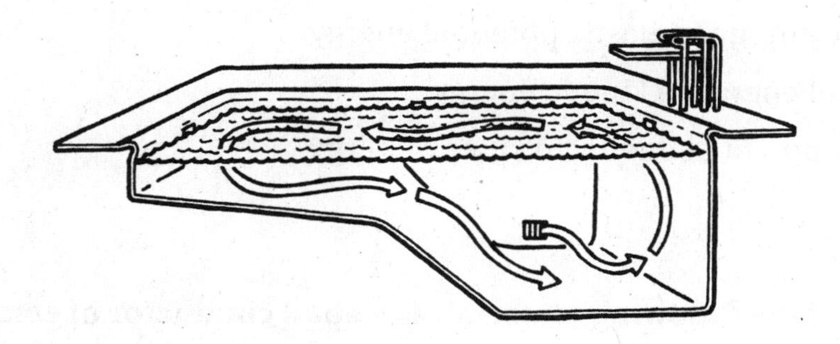

A density

B ductility

C inertia

D weight

74. **How is bioengineering applied to the development of new drugs?**

F making advertisements that tell doctors about the drugs

G finding plants that contain natural drugs

H reducing costs by building a larger manufacturing plant

J making drugs based on knowledge about the chemical reactions in a cell

75. What type of energy transformation is represented by a bear bending a small tree to reach fruit in its branches?

A kinetic energy to gravitational potential energy

B kinetic energy to elastic potential energy

C thermal energy to kinetic energy

D elastic potential energy to kinetic energy

76. Which of the following materials is a good conductor of electricity?

F glass

G air

H mercury

J all of the above

77. Which of the following could be an intended benefit of a biodegradable plastic grocery bag?

A increased cost to the store

B decreased pollution

C decreased strength of the plastic

D keeps food cold

1. **The graph below shows the monthly distribution of rainfall as compared to the average rainfall. Use the graph to answer the question that follows.**

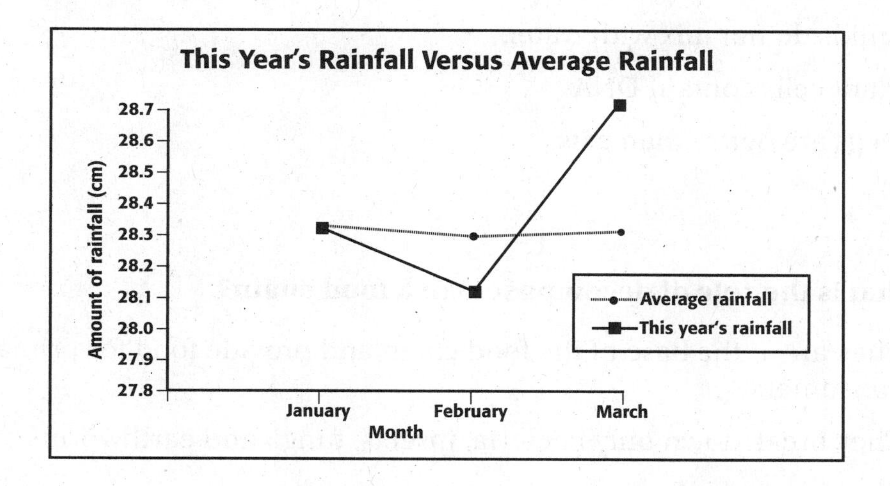

Why is this graph biased?

- **A** The distribution of rain on the y-axis does not reflect meaningful gradations.
- **B** The distribution of time on the x-axis does not reflect meaningful gradations.
- **C** The graph does not distinguish between heavy and light rainfall.
- **D** The graph does not say how the amount of rain was measured.

2. **Which part of a circuit changes electrical energy into another form of energy?**

- **F** energy source
- **G** switch
- **H** wire
- **J** load

3. Which of the following is not an example of a hypothesis that can be tested with an experiment?

A Living things grow and develop.

B Lipids do not mix with water.

C Plant cells contain DNA.

D Dogs are better than cats.

4. What is the role of decomposers in a food chain?

F They are at the base of the food chain and provide food for primary consumers.

G They break down only bacteria, insects, fungi, and earthworms.

H They break down waste and rotting matter.

J They provide food and shelter for animals.

5. In the figure below, which bulb burning out would mean that no current could flow through the circuit?

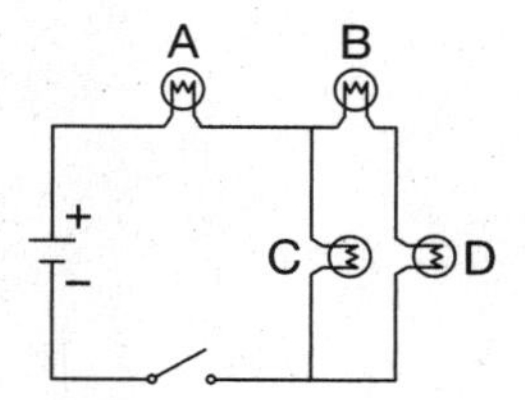

A bulb A

B bulb B

C bulb C

D bulb D

6. What happens to the energy that is <u>not</u> used when an engine is less than 100% efficient?

F It is destroyed during combustion.

G It is used to decrease entropy.

H It is converted to heat.

J It is converted to matter.

7. **A preference for a certain outcome over another is**

A bias.

B scientific inquiry.

C random sampling.

D controlled experimentation.

8. **In an ecosystem, wolves eat elk, and elk eat grass. If the number of wolves decreases, what will likely occur first?**

F The number of elk will increase.

G The number of elk will decrease.

H The amount of grass will decrease.

J The amount of grass will increase.

9. **Constance made the table below during a field investigation on ocean life. Which zone would an organism most likely be found in if it has no eyes and can live in water as hot as 80°C?**

Characteristics of Several Ocean Zones

Zones	Description
Intertidal	air, sun, and water exposure; crashing waves
Neritic	water depth less than 200 m; lots of sunlight; relatively warm water
Benthic	very deep water; no light; cold except near thermal vents that emit heat and chemicals

A neritic zone

B benthic zone

C intertidal zone

D intertidal or neritic zones

10. **In what part of the Milky Way galaxy is our solar system located?**

F near the center

G above the central axis

H in a spiral arm between the center and edge of the galaxy

J at the outside edge of the spiral arms

11. What part of Earth receives the most direct solar energy?

A the poles

B the equator

C midway between the poles and the equator

D equal at all points on Earth's surface

12. Which of the following materials is an electrical insulator?

F copper

G iron

H aluminum

J rubber

13. **Which regions in the map below receive the greatest annual precipitation and have the most diverse animal life?**

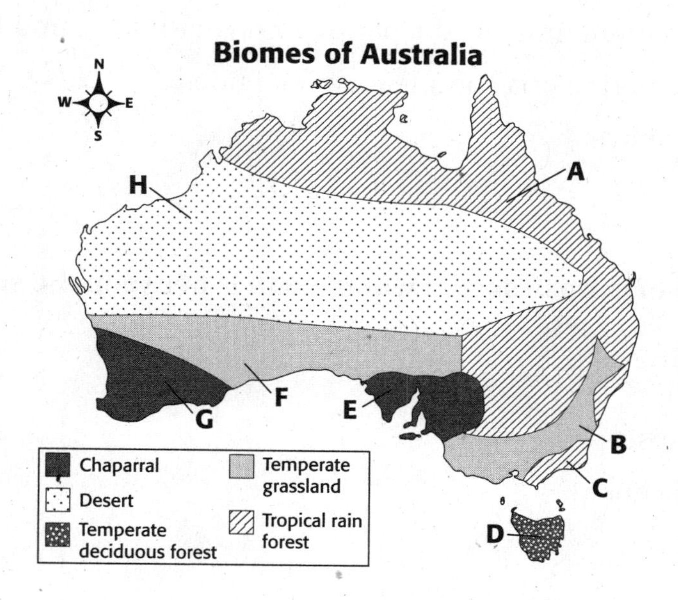

A Regions *A* and *C*

B Region *B* and *F*

C Region *E* and *G*

D Region *D* and *H*

14. **Which of the following steps would <u>not</u> be part of a scientific investigation?**

F collecting evidence through experiments

G revising data to fit the hypothesis

H analyzing measurements

J drawing conclusions from the data

15. The phases of the moon occur about every 28 days because

A the period of rotation of the moon is the same as its period of revolution.

B 28 days is approximately the period of revolution around Earth.

C that is the period of a complete cycle of tides.

D all of the above

16. What type of eclipse occurs when Earth is between the sun and the moon?

F solar eclipse

G lunar eclipse

H annular eclipse

J total eclipse

17. Which of the following food chains represents how energy flows in the food web shown below?

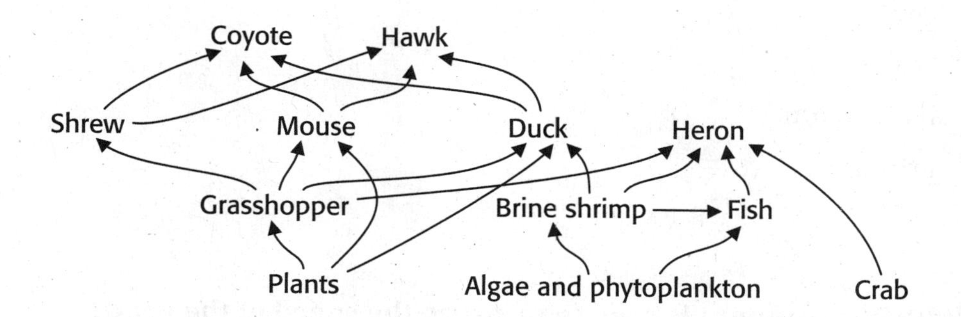

A coyote → hawk → duck → heron

B algae → brine shrimp → duck → hawk

C plants → grasshopper → duck → heron

D hawk → mouse → grasshopper → plants

18. What beneficial result might happen if a scientist discovers her hypothesis is incorrect?

F Her results might lead to another hypothesis.

G Other scientists might repeat her work.

H She could keep her results secret.

J She might give up.

19. A surface current is a horizontal movement of ocean water at or near the ocean's surface. Surface currents are caused by

A waves.

B winds.

C a jet stream.

D a rip tide.

20. Which instrument is used to measure the speed of the wind?

F barometer

G anemometer

H wind vane

J radar

21. The term abiotic refers to

A all nonliving factors in an environment.

B adaptations an organism makes to survive.

C a type of tissue present in plants.

D a type of tissue present in animals.

22. Earth completes one full rotation on its axis every 24 hours. During this period of time, a person living in Tennessee experiences

F 24 hours of darkness.

G 24 hours of daylight.

H a period of daylight and a period of darkness.

J a lunar eclipse.

23. Anchorage, Alaska, will experience its coldest temperatures for the year

A when Earth is farthest from the sun.

B during the equinoxes.

C when the Northern Hemisphere is tilted away from the sun.

D during the summer solstice.

24. Which of these factors can limit the size of a population?

F temperature

G light

H soil composition

J all of the above

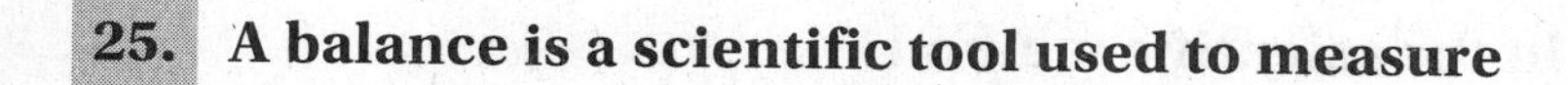

25. A balance is a scientific tool used to measure

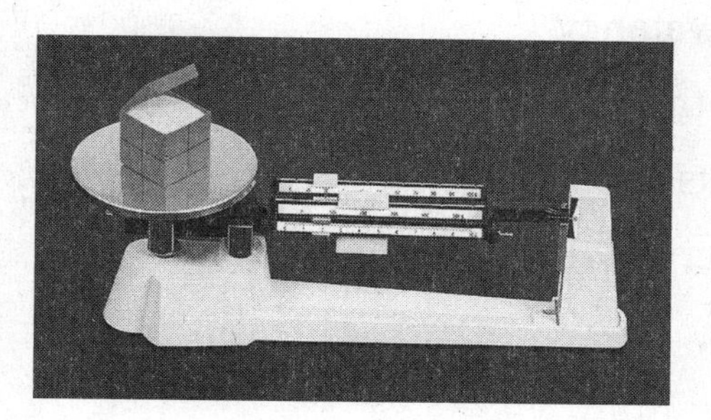

A temperature.

B time..

C volume.

D mass.

26. A dock was built over a large bed of sea grass in a manatee habitat. The dock shaded the bed of sea grass from the sun. The population of manatees decreased in the area even though the manatees could still swim under the dock. Why did the population of manatees decrease?

F The sea grass was poisoned.

G The sea grass grew too thick.

H The dock kept the manatees from reaching the sea grass.

J The sea grass died because the dock shaded it from the sun.

27. Which is not a step in the scientific method?

A finding an equivalency

B forming a hypothesis

C analyzing results

D making an observation

28. The volume of a solid is measured in

F liters.

G grams.

H cubic centimeters.

J all of the above.

29. **What type of front is represented in the illustration below?**

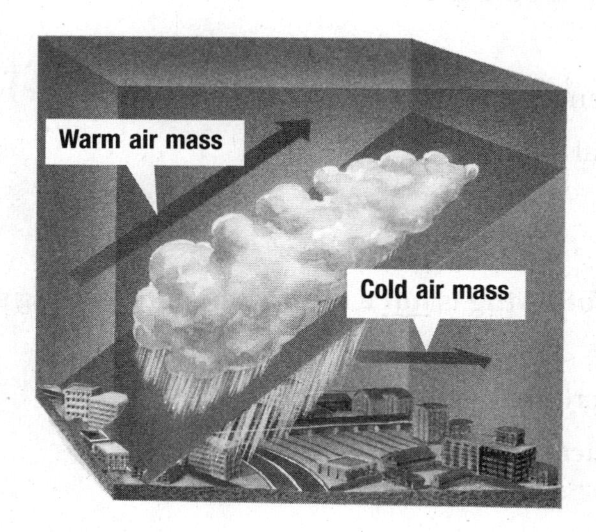

A cold front

B warm front

C stationary front

D occluded front

30. **Two objects at different temperatures are in contact. Which of the following happens to their thermal energy?**

F Their thermal energies remain the same.

G Thermal energy passes from the cooler object to the warmer object.

H Thermal energy passes from the warmer object to the cooler object.

J Thermal energy passes back and forth equally between the two objects.

31. What determines an object's thermal energy?

A the motion of its particles

B its size

C its potential energy

D its mechanical energy

32. Which of the following is <u>not</u> a requirement of writing science instructions?

F accurate description

G reliance on facts

H logical flow

J creative wording

33. **The graph below shows the average temperature (line) and monthly precipitation for a particular area. What biome would you find in this area?**

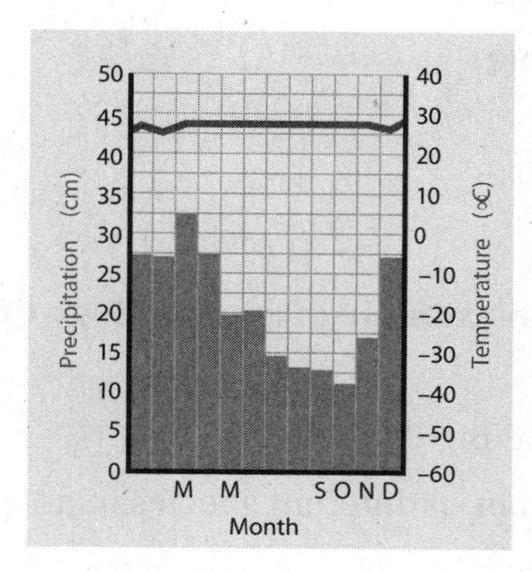

A temperate forest

B tropical desert

C tundra

D tropical rain forest

34. **Which technology was developed by scientists and engineers who studied the properties of charged particles?**

F new types of ceramic glazes

G internal combustion engine

H transmission of electrical energy

J better alloys for airplane engines

35. Which of the following is not a type of meteorite?

A stony meteorite

B stony-iron meteorite

C rocky-iron meteorite

D metallic meteorite

36. Which of these is an example of a technology that responds to a social need?

F development of the Big Bang theory

G counting the number of different species in an ecosystem

H comparing the electrical conductivity of copper and iron

J development of a new model to improve traffic flow in a congested city

37. How do surface currents affect circulation in the atmosphere?

A The motion o the water drives the wind.

B Currents bring warmer or cooler water that affects air temperature.

C Moving water absorbs more air than still water.

D Currents generate large waves that disrupt air flow.

38. By which method does most thermal energy in the atmosphere circulate?

F conduction

G convection

H condensation

J transpiration

39. Which of these factors would not be an important consideration during the design of a new kind of ship?

A the cost of materials

B the name of the new ship

C the type of engine that will be used to provide power

D the type of cargo that the ship will carry

40. Which of these is a kind of potential energy?

F thermal energy

G chemical energy

H kinetic energy

J sound energy

41. What happens to the majority of the sun's energy that reaches Earth's atmosphere?

A It is absorbed by the land and ocean.

B It is reflected by clouds and air.

C It is reflected by Earth's surface.

D It is absorbed by ozone, clouds, and atmospheric gases.

42. An engineer is designing a new kind of bridge to cross a wide river. Which of the steps of the design process would occur earliest in the design process?

F decide how strong the cables need to be

G start manufacturing the deck for the bridge

H build a model of the bridge

J determine the length of the bridge needed to cross the river

43. Which of these could be an unintended consequence of engineering a crop to be toxic to an insect that eats it?

A increased production of the crop

B higher profits for the farmer

C transfer of the resistance to related weeds

D other harmful insects are also killed when they eat the plant

44. Which of these prevailing winds would carry the <u>most</u> moisture?

F winds formed from warm air

G winds formed from cold air

H winds formed over land

J winds formed near mountains

45. Herbivores, carnivores, and scavengers are all examples of

A producers.

B decomposers.

C consumers.

D omnivores.

46. A limiting factor is a resource that

F keeps the population from moving.

G causes the extinction of a population.

H keeps the size of a population from growing too quickly.

J provides an adaptive strategy for a population.

47. Which the following sources of information about the safety of a child's car seat is the <u>most</u> reliable?

A an advertising brochure from the manufacturer

B a statement from a doctor that the seat provides the best protection

C an article from a government testing laboratory comparing brands of car seats

D information on the box in which the car seat is packaged

48. Which of these products is <u>likely</u> to have been developed by a bioengineer?

F artificial leg

G automobile tire

H airplane propeller

J all of the above

49. How long does it take the sun's energy to reach Earth?

A about 8 minutes

B about 8 hours

C about 80 hours

D about 8 days

50. Nature's recyclers are

F predators

G decomposers

H producers

J omnivores

51. **Which alignment describes the positions of the sun, the moon, and Earth during a new moon?**

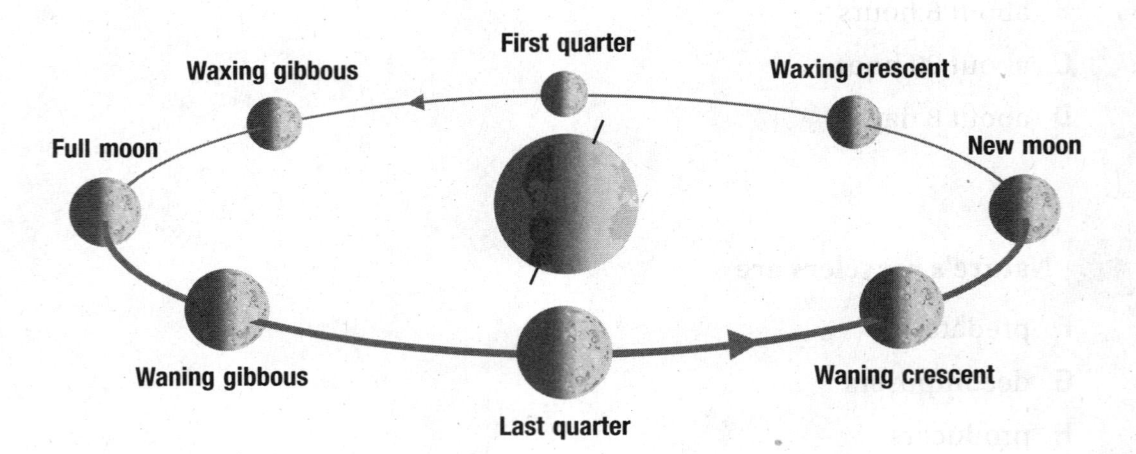

A The moon is located between the sun and Earth.

B The Earth is located between the sun and the moon.

C The moon is located 90° from a line between the sun and Earth.

D The Earth is located 90° from a line between the sun and the moon.

52. **The maximum number of hours of daylight for Sydney, Australia occurs on or around**

F March 21.

G December 21.

H September 22.

J June 21.

53. The sun's rays strike Earth's surface at different angles in different places. This phenomenon occurs because

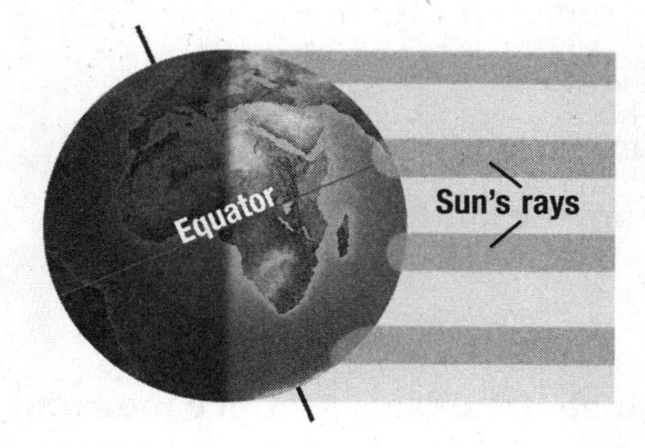

A Earth is tilted.

B Earth's surface is curved.

C Earth is the correct distance from the sun.

D Earth's surface is rough and uneven.

54. Which of these organisms is located at the base of the energy pyramid?

F oak tree

G bat

H squirrel

J earthworm

55. Which of the following conditions contributes to the decrease in a hurricane's strength as it moves from the ocean onto land?

- **A** the lack of warm, moist air over land
- **B** the uneven land surface
- **C** rising hot air from the land
- **D** sinking hot air from the land

56. The process by which organisms produce food and oxygen from sunlight is called

- **F** consumption.
- **G** photosynthesis.
- **H** respiration.
- **J** absorption.

57. In the Northern Hemisphere, winter begins on or around the winter solstice. This date is

A March 21.

B December 21.

C September 22.

D June 21.

58. The climate of Iceland is milder than that of Greenland because

F Greenland is farther north.

G the Gulf Stream carries warm water to Iceland but not Greenland.

H wind currents carry tropical air to Iceland but not Greenland.

J Iceland is closer to Europe.

59. If squirrels in a forest begin to die because of ingesting pesticides, what may happen to the ecosystem?

A The organisms in the ecosystem will all be affected in some way.

B The ecosystem will remain unchanged.

C The organisms that ate squirrels will reproduce faster.

D The ecosystem will continue to thrive.

60. Which of the following are driven by energy from the sun?

F the water cycle

G ocean currents

H winds

J all of the above

61. An example of a biotic factor in a freshwater lake is

A the amount of sunlight that passes through the water.

B the temperature of the water.

C the depth of the water.

D the fish that live in the lake.

62. Two different populations rely on the same resource for their food supply. The populations are said to be in

F the same species.

G symbiosis.

H competition.

J the food chain.

63. Oceans help to moderate the temperatures of lands around them because they

A keep surface currents from moving too close to the land.

B lead to heavy snowfall over the water instead of the land.

C cause strong winds near the surface.

D absorb and release heat more slowly than the land does.

64. Which location below is likely to experience the largest temperature difference between summer and winter based on the amount of sunlight received?

F Alberta, Canada

G Memphis, Tennessee

H Key West, Florida

J Mexico City, Mexico

65. This map shows the location of low- and high-pressure belts across North America and South America. What causes the formation of the high-pressure belt?

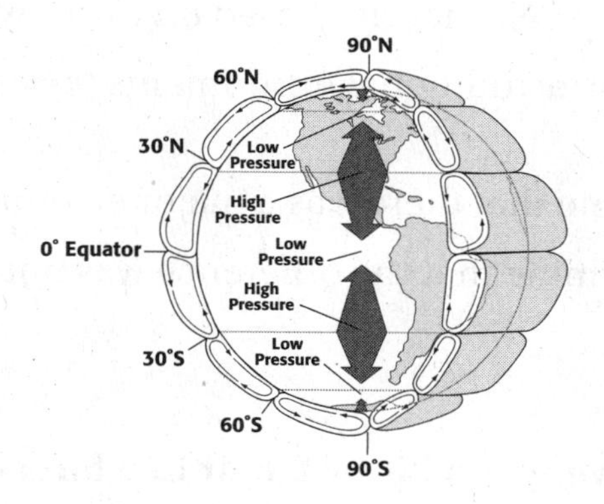

A the sinking and warming of cool air

B the rising and warming of cool air

C the rising and cooling of warm air

D the sinking and cooling of warm air

66. If a location has few seasonal changes in its temperature or the length of its days, the location is

F in the northern temperate zones.

G at the poles.

H near the equator.

J in the southern temperate zones.

67. If an Earth scientist presented data at a conference that the planet is actually about one billion years older than current theories suggest, how would other scientists most likely react?

A They would revise their theories based on the new evidence.

B They would perform additional experiments to see if the result can be replicated.

C They would assume that there was a major error in the experiment.

D They would determine that the difference was not important.

68. Which of the following is a biotic factor in a forest ecosystem?

F The rock formations that make up the forest floor.

G The mosses that cover sides of tree trunks.

H The average temperature in the winter.

J The amount of sunlight that reaches the forest floor.

69. Which regions in the map below receive the greatest annual precipitation and have the most diverse animal life?

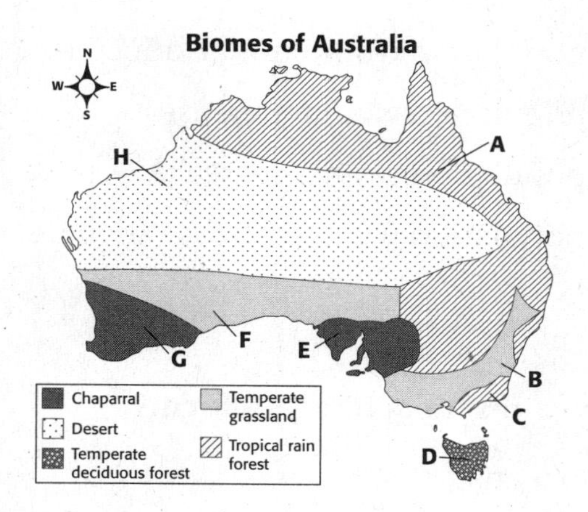

A Region A and C

B Region B and F

C Region E and G

D Region D and H

70. Resources such as water, food, or sunlight are likely to be limiting factors

F when population size is decreasing.

G when predators eat their prey.

H when the population is small.

J when a population is approaching the carrying capacity.

71. Imagine that the herbivore population in an ecosystem decreases for some reason. How would the population of the carnivores in that ecosystem be affected?

A The carnivore population would not be affected.

B The carnivore population would increase.

C The carnivore population would decrease.

D The carnivore population would increase then decrease.

72. Where are hurricanes most likely to form?

F above tropical oceans

G above temperate oceans

H above tropical landmasses

J above temperate landmasses

73. A characteristic of every energy transfer is that

A thermal energy is transferred.

B convection usually occurs.

C some energy is lost.

D conservation saves all the energy.

74. After high-pressure areas are created around the poles, cold polar air flows toward

F the equator.

G the North Pole.

H the South Pole.

J the atmosphere.

75. Why did the elk population of the United States increase when the gray wolf population decreased?

- **A** Fewer gray wolves ate grass that the elk grazed on.
- **B** Not enough gray wolves were around to control the elk population.
- **C** More places were available for elk to live.
- **D** The elk were able to eat the animals that the gray wolves usually preyed on.

76. Which of the following is not part of the biosphere?

- **F** oceans
- **G** the atmosphere
- **H** mountain ranges
- **J** the moon

77. Two scientists in a laboratory collect data from an experiment. Using the same results, they develop two hypotheses that are very different. What can you conclude based on this information?

- **F** One hypothesis is correct, and the other is incorrect.
- **G** Both scientists based their conclusions on faulty data.
- **H** There was an error in the analysis of the data.
- **J** More experiments are needed to answer the questions raised by the data.

Notes

Notes

Notes

Notes

Notes

Notes